AF896501

**Unmanned System Technologies**

Springer's Unmanned System Technologies (UST) book series publishes the latest developments in unmanned vehicles and platforms in a timely manner, with the highest of quality, and written and edited by leaders in the field. The aim is to provide an effective platform to global researchers in the field to exchange their research findings and ideas. The series covers all the main branches of unmanned systems and technologies, both theoretical and applied, including but not limited to:

- Unmanned aerial vehicles, unmanned ground vehicles and unmanned ships, and all unmanned systems related research in:
- Robotics Design
- Artificial Intelligence
- Guidance, Navigation and Control
- Signal Processing
- Circuit and Systems
- Mechatronics
- Big Data
- Intelligent Computing and Communication
- Advanced Materials and Engineering

The publication types of the series are monographs, professional books, graduate textbooks, and edited volumes.

Ishtiaq Ahmad · Ramsha Narmeen ·
Zeeshan Kaleem · Zdenek Becvar ·
Hina Tabassum · Ahmed Alkhayyat
Editors

# Reconfigurable Intelligent Surfaces for Unmanned Aerial Vehicle Communication

Springer

*Editors*
Ishtiaq Ahmad
Department of Telecommunication, Engineering
Faculty of Electrical Engineering
Czech Technical University in Prague
Prague, Czech Republic

Zeeshan Kaleem
Department of Computer Engineering and Interdisciplinary Research Center for Smart Mobility and Logistics
King Fahd University of Petroleum and Minerals (KFUPM)
Dhahran, Saudi Arabia

Hina Tabassum
Lassonde School of Engineering
York University
Toronto, Canada

Ramsha Narmeen
Department of Telecommunication
Faculty of Electrical Engineering
Czech Technical University in Prague
Prague, Czech Republic

Zdenek Becvar
Faculty of Electrical Engineering
Czech Technical University in Prague
Prague, Czech Republic

Ahmed Alkhayyat
College of Technical Engineering
The Islamic University
Najaf, Iraq

ISSN 2523-3734 ISSN 2523-3742 (electronic)
Unmanned System Technologies
ISBN 978-981-96-9661-1 ISBN 978-981-96-9662-8 (eBook)
https://doi.org/10.1007/978-981-96-9662-8

© The Editor(s) (if applicable) and The Author(s), under exclusive license to Springer Nature Singapore Pte Ltd. 2026

This work is subject to copyright. All rights are solely and exclusively licensed by the Publisher, whether the whole or part of the material is concerned, specifically the rights of translation, reprinting, reuse of illustrations, recitation, broadcasting, reproduction on microfilms or in any other physical way, and transmission or information storage and retrieval, electronic adaptation, computer software, or by similar or dissimilar methodology now known or hereafter developed.
The use of general descriptive names, registered names, trademarks, service marks, etc. in this publication does not imply, even in the absence of a specific statement, that such names are exempt from the relevant protective laws and regulations and therefore free for general use.
The publisher, the authors and the editors are safe to assume that the advice and information in this book are believed to be true and accurate at the date of publication. Neither the publisher nor the authors or the editors give a warranty, expressed or implied, with respect to the material contained herein or for any errors or omissions that may have been made. The publisher remains neutral with regard to jurisdictional claims in published maps and institutional affiliations.

This Springer imprint is published by the registered company Springer Nature Singapore Pte Ltd.
The registered company address is: 152 Beach Road, #21-01/04 Gateway East, Singapore 189721, Singapore

If disposing of this product, please recycle the paper.

# Acknowledgement

The author (Zeeshan Kaleem) would like to acknowledge the support received from King Fahd University of Petroleum & Minerals (KFUPM) during the writing phase of the book

# Introduction

This book offers a comprehensive exploration of use cases, advanced intelligent technologies, and layered architectures for RIS-assisted UAV communications, with a focus on both standard non-intelligent and intelligent approaches tailored to these systems. Recent advancements in beamforming and signal reflection for RIS-assisted UAV networks have attracted significant attention from researchers and academics due to their ability to dynamically enhance communication coverage, quality, and privacy by reflecting incoming signals. Furthermore, the integration of machine learning techniques into RIS-assisted UAV communications introduces intelligence, aligning with the goals of sustainable, eco-friendly, and self-sustaining 6G networks. This innovative approach positions RIS-assisted UAV systems as a promising solution to overcome the limitations of current wireless technologies.

Chapter 1 reviews recent developments in unmanned aerial vehicles (UAVs) and reconfigurable intelligent surfaces (RIS), both of which have shown promising applications through the use of intelligent technologies and layered architectures. Extensive research has highlighted significant improvements in wireless network performance driven by both RIS and UAVs individually. By integrating RIS into UAV platforms, communication capabilities are enhanced through the inclusion of adaptable surfaces that optimize signal strength and reduce interference in various environments. However, the integration of these two technologies presents challenges, as conventional optimization methods are insufficient to address their complexities. As a result, there is an increasing need for real-time optimization, artificial intelligence-driven solutions, and the adoption of emerging techniques to effectively tackle these challenges. In response, the study outlines key technologies and research directions for RIS-assisted UAV communication.

Chapter 2 aims to enhance the communication capabilities of UAV-borne IoT platforms in the presence of jamming threats by leveraging AI, particularly Deep Reinforcement Learning (DRL) algorithms like DDPG and TD3. The UAV, acting as the agent, optimizes both its flight path and RIS (Reconfigurable Intelligent Surface) configuration to maximize data rates while mitigating jamming effects. By using the UAV's received data rate as input, the system adapts to jamming conditions without the need for complex instantaneous CSI estimation. This AI-based method also addresses challenges in RIS-assisted jamming rejection by allowing the UAV to dynamically adjust the wireless environment, improving signal quality. Additionally, the system supports UAVs in radio surveillance tasks by capturing high-resolution radio fingerprints of suspicious transmitters. The DRL framework enables efficient surveillance and eavesdropping by learning from observed transmit rates, bypassing the need for intricate CSI estimations and overcoming mobility and environmental constraints.

Chapter 3 focuses on the integration of Unmanned Aerial Vehicles (UAVs) with Reconfigurable Intelligent Surfaces (RIS) to enhance wireless communication networks, particularly for beyond 5G and future 6G applications. UAVs offer flexibility and cost-effectiveness, making them ideal for a variety of uses such as aerial inspection, drone photography, crop monitoring, delivery services, and wireless communication. When combined with RIS, UAVs can significantly improve network coverage, reliability, and energy efficiency by adjusting signal propagation to reduce interference and optimize communication. The primary objective of this chapter is to enhance system performance by jointly optimizing UAV altitudes and RIS elements. To achieve this, a novel optimization approach is introduced, combining Riemannian Gradient and Particle Swarm Optimization (RGO) techniques, which are applied through an alternating optimization strategy to address the complex problem. Experimental results validate the superiority of the proposed method over traditional strategies, showcasing its potential in improving UAV-RIS assisted network performance.

Chapter 4 focuses on enhancing the security of UAV communication networks through RIS-assisted transmission schemes. UAVs face inherent vulnerabilities to eavesdropping due to the high probability of line-of-sight channels. The integration of Reconfigurable Intelligent Surfaces (RIS) offers a promising solution by manipulating the propagation environment to improve security. Specifically, we address an optimization problem aimed at maximizing the average secrecy rate, combining transmit beamforming, passive beamforming, and UAV trajectory optimization. To solve this, we decompose the problem into three sub-problems and propose an iterative algorithm for their alternating resolution. RIS plays a crucial role in strengthening the security framework by enhancing signal quality for intended users while degrading eavesdropping links. The effectiveness of our approach is demonstrated through simulations, providing valuable insights. Finally, we identify challenges that need to be addressed for practical implementation in this rapidly evolving field.

Chapter 5 Systems explores an unmanned aerial vehicle (UAV)-assisted, reconfigurable intelligent surface (RIS)-enhanced uplink non-orthogonal multiple access (NOMA) communication system, where user equipment (UEs) aims to transmit confidential information to the UAV while an eavesdropper is present. We analyze the impact of transmit power on the secrecy rate under Rayleigh fading channels, considering both continuous and discrete phase shifts at the RIS. Simulation results demonstrate that increasing the UEs' transmit power enhances the secrecy rate, while the use of discrete phase shifts at the RIS degrades system performance.

Chapter 6 aims to optimize energy efficiency in a wireless communication model involving an RIS-assisted UAV base station serving multiple ground users. UAVs provide broader coverage than ground-based stations and typically maintain a line-of-sight (LOS) link with users, minimizing propagation losses and transmission power. However, obstacles can disrupt LOS, leading to increased propagation losses and power requirements. To address this, reconfigurable intelligent surfaces (RIS) are used to redirect the UAV's signal, reducing transmission power. The chapter focuses on the challenge of communication mode selection and power allocation, with the goal of maximizing energy efficiency by optimizing bits per joule. A mathematical formulation is presented, and numerical simulations are conducted to assess the solution's performance.

Chapter 7 explores the integration of Reconfigurable Intelligent Surfaces (RIS) with Unmanned Aerial Vehicles (UAVs) to enhance wireless connectivity and network optimization, particularly for 6G and beyond. The combination of RIS and UAVs offers practical solutions to complex challenges in communication and data delivery, providing dynamic trajectory planning to create Line-of-Sight (LoS) links in Non-Line-of-Sight (NLoS) areas. This synergy optimizes resource allocation, spectrum sharing, and network performance. The chapter presents two case studies: the first demonstrates the use of tethered UAVs with RIS to restore failed base station fronthaul links, while the second introduces ASTAR-RIS, which actively amplifies signals to improve the signal-to-noise ratio (SNR) for remote users. Simulations show that the ASTAR-UAV model significantly enhances network performance and resource allocation, highlighting its potential to revolutionize UAV-assisted wireless networks in the 6G era.

Chapter 8 focuses on evaluating the performance of UAV communication systems using diversity combining techniques, particularly in relation to Bit Error Rate (BER) in scenarios involving Rayleigh and Ricean fading models. For years, these diversity combinations, including selective combining, maximal ratio, equal gain, and direct combining, have been used to address BER challenges. While previous models based on Rayleigh and Ricean schemes have been applied in smaller-scale UAV communication analysis, they have not fully depicted the most reliable and effective models for UAV systems. This study provides a comprehensive survey of these models and the associated issues in UAV communication, offering a simplified mathematical expression to compare the two fading models. The simulated results show that the Rayleigh model outperforms the Ricean model in terms of BER performance.

Chapter 9 explores methods to secure Unmanned Aerial Vehicle-Assisted Reconfigurable Intelligent Surface (UAV-RIS) systems in next-generation communication networks. The study proposes using an advanced deep machine learning framework called Long Short-Term Memory Deep Deterministic Policy Gradient with Prioritized Experience Replay (LSTM-DDPG-PER) to address security challenges and ensure reliable communication in UAV-assisted RIS networks by mitigating potential threats. The LSTM-DDPG-PER framework enhances learning efficiency, convergence speed, and resilience to adversarial attacks by frequently revisiting key experiences. Through simulations, the study demonstrates that combining UAV-RIS with LSTM-DDPG-PER significantly outperforms traditional methods in countering attacks on UAV-RIS communication.

# Contents

# Chapter 1
# Reconfigurable Intelligent Surface for Unmanned Air Vehicle Communications: Emerging Technologies and Research Directions

**Ishtiaq Ahmad and Ramsha Narmeen**

**Abstract** In recent developments, unmanned aerial vehicles (UAVs) and reconfigurable intelligent surfaces (RIS) have emerged with promising applications, employing intelligent technologies and layered architectures. Extensive research has demonstrated individual advancements in wireless network performance by both RIS and UAVs. Integrating RIS into UAV platforms enhances communication capabilities by incorporating adaptable surfaces, optimizing signal strength, and mitigating interference in diverse environments. However, challenges arise in seamlessly integrating both technologies, as conventional optimization solutions fall short in addressing these complexities. Consequently, there is a growing imperative for real-time optimization, artificial intelligence-assisted solutions, and the integration of developing techniques to overcome such obstacles effectively. In response to these challenges, this study delineates crucial technologies and research directions relevant to RIS-assisted UAV communication.

**Keywords** Intelligent UAVs · RIS · Machine learning

## 1.1 RIS Communications and Standards

The unprecedented constraints for extensive, high-quality wireless services present prominent challenges to recent cellular networks. Designing a fifth-generation (5G) communication framework aims at applications such as massive machine-type communications (mMTC), rate-centric enhanced mobile broadband (eMBB), and ultra-

I. Ahmad · R. Narmeen (✉)
Faculty of Electrical Engineering, Czech Technical University in Prague, Prague, Czech Republic
e-mail: ramshanarmeen1992@gmail.com; narmeram@fel.cvut.cz

I. Ahmad
e-mail: ishtiaq001@gmail.com; ahmadish@fel.cvut.cz

© The Author(s), under exclusive license to Springer Nature Singapore Pte Ltd. 2026

I. Ahmad et al. (eds.), *Reconfigurable Intelligent Surfaces for Unmanned Aerial Vehicle Communication*, Unmanned System Technologies,
https://doi.org/10.1007/978-981-96-9662-8_1

reliable low-latency communications (URLLC). The objectives of sixth-generation (6G) wireless communication framework are anticipated to be transformative and innovative, involving applications like instantaneous, data-driven, ultra-massive, and ubiquitous wireless connections, along with linked intelligence [1, 2]. Hence, novel transmission techniques are required to obtain a system that accommodates such emerging applications and services. Recently, reconfigurable intelligent surfaces (RISs), alternatively referred to as intelligent reflecting surfaces (IRSs) [3, 4] or large intelligent surfaces (LISs) [5], have been highly investigated.

RISs are recognized as a key technology demonstrating substantial advantages in spectrum and energy efficiency, as well as cost-effectiveness. RIS emerges as a fitting solution that complements existing cutting-edge network and communication technologies. Particularly in massive multiple-input multiple-output (MIMO) systems, integrating RIS presents a promising strategy to address the significant hurdles posed by hardware costs and energy consumption associated with extending the coverage of a network.

RIS involves the use of specially designed materials or surfaces that can dynamically alter their electromagnetic properties to control and manipulate radio frequency signals. Typical RIS constitutes a programmable two-dimensional framework with numerous elements designed to redirect signals toward receivers without amplification. The finely adjustable RIS elements enable precise control of radio propagation for various purposes. The optimization of RIS phase shifts is frequently undertaken in conjunction with factors like the transmit power of the RIS, beamforming design, and placement of the RIS. RISs have the ability to enhance signal strength, improve coverage, and optimize communication links.

### *1.1.1 Transitioning from Physics to Wireless Communications: The Role of RIS*

RISs can be implemented using either metamaterial or patch-array-based techniques considering the structures of RIS. Metamaterial-based RISs are commonly known as metasurfaces. Positioned at various spots, RIS is able to be engineered to function as reflecting or refracting surfaces linking the base station (BS) and the user equipment (UE) or as waveguide surfaces operational at the BS. Regarding tuning processes, RISs can undergo electrical, mechanical, or thermal reconfiguration. Based on their energy consumption, RISs fall into categories, such as passive-lossless, passive-lossy, or active, with their passive or active nature influencing their critical performance capacities. It is important to note that RISs is not able to be entirely passive due to the fundamental configurability. In this context, we explore three crucial RIS operational modes: waveguide [6], refraction [7], and reflection [8]. Under the three operational states, the RIS transforms and emits a wave (that is either prompted by an incident wave or supplied through a waveguide) into the intended propagating wave in open space.

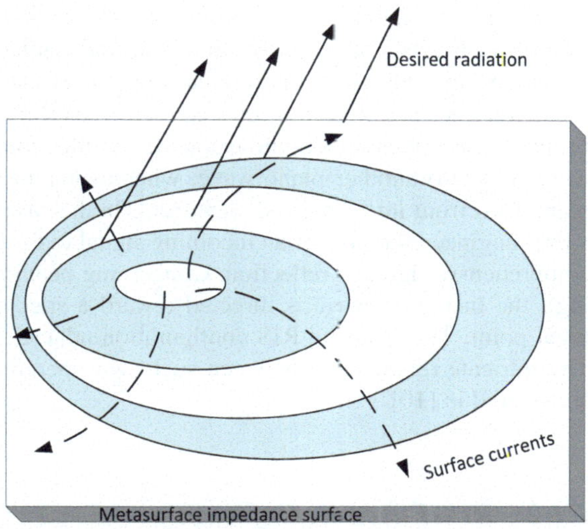


**Fig. 1.1** Conceptual illustration of pre-designed holographic waveguide that is based RIS

The components within the metasurface of RIS are conceptualized as independent magnetic dipoles. Adjusting the polarizability allows the antenna of the metasurface to execute beamforming. Every element on the metasurface functions as a micro-antenna. In contrast to traditional antenna arrangements, the condensed metasurface waveguide requires less room and is able to transmit over broader directions.

A wireless signal can be described as an electromagnetic (EM) wave moving through a three-dimensional space. Signal strength experiences attenuation or decrease as the electromagnetic wave travels via space and mingles with scattering objects. According to the fundamental assumptions of electromagnetism, the strength of the signal per unit area is directly proportional to the square of the wave's electric field in a current medium. In the context of refractive and reflective smart surfaces, comprehending how EM waves interrelate with neighboring objects becomes essential.

Regarding waveguide-founded RISs, the operational principle is succinctly outlined below: In [9], the manipulation of electromagnetic waves by the metasurface provides a connection between the wave's three-dimensional free space and two-dimensional surface. Consequently, the metasurface is likened to a hologram, carrying supplementary data regarding the propagated wave in three-dimensional (3D) space. Once stimulated by the source, such pre-arranged data integrates into the radiated field. Figure 1.1 abstractly illustrates a pre-designed holographic waveguide-based RIS.

The EM features of RIS, like phase incoherence, undergo reconfiguration via adjusting the surface resistance using numerous procedures. In addition to electrical power, alternative methods such as thermal innervation, optical pumping, and physical expansion can be employed. Surrounded by these, electrical control stands out as the favorably appropriate option, as the electrical voltage is easily quantifiable and manageable through field-programmable gate array (FPGA) chips. The selection of substances for RIS encompasses graphene and semiconductors.

Among the numerous operational operations and setups of RISs, inconsistent reflection and beamforming are extensively exploited in wireless infrastructures. From a wave-optics standpoint, inconsistent reflection involves transforming a plane wavefront into another plane wave, whereas beamforming entails transforming a plane wavefront into a desired wavefront. In the case of inconsistent reflection, the RIS is engineered to mirror an incoming signal to far-field locations, adhering to the comprehensive laws of reflection. Concerning beamforming (also known as focusing), the incoming wave is directed toward a specific area, generally termed the focal point. The requisite RIS configuration adheres to the co-phase specification. The intricate relationship between such two operating philosophies is thoroughly considered in [10].

### 1.1.2 RIS-Based Beamforming

Implementing RISs leads to significant improvements in the performance of wireless networks. Considering this, we can say that the advantages of employing RIS in smart radio systems are undeniable. Firstly, RISs can be ubiquitously deployed to harness EM waves that would otherwise dissipate in space. Utilizing reconfigurable electromagnetic materials, objects in the environment, such as building facades, ceilings, furniture, and clothing, can be coated with RISs. Secondly, RISs contribute to environmental sustainability by aligning with the principles of green communications. As we know, RISs are practically passive; they do not consume additional energy compared to traditional wireless networks. Thirdly, RISs facilitate full-band and full-duplex communication by reflecting electromagnetic waves only. Fourthly, the power gain of a RIS adheres to the quadratic scaling principle, differing from the linear power scaling principle of a traditional active antenna matrix.

RIS in wireless communication is commonly employed for two primary objectives: (1) functioning as an RF chain-free transmitter and (2) serving as a passive beamforming surface that enhances the incoming signal (generally from a BS) and reflects it to the targeted UE, referred to as an IRS. In Fig. 1.2, we can see that an RIS is employed to aid the communication from the BS to the UEs by passively reflecting the signals. Recently, combined active and passive beamformers have been developed aimed at minimizing transmission power at the BS, utilizing continuous as well as discrete phase shifts [11, 12], respectively. Additionally, a passive beamformer is optimized for maximizing the ergodic sum rate and energy efficiency in [13, 14], respectively. Furthermore, a theoretical analysis of the average symbol error rate (SER) arising from a passive beamformer is also thoroughly examined [15]. It demonstrated that SER undergoes an exponential decay as the number of reflecting elements on the RIS increases.

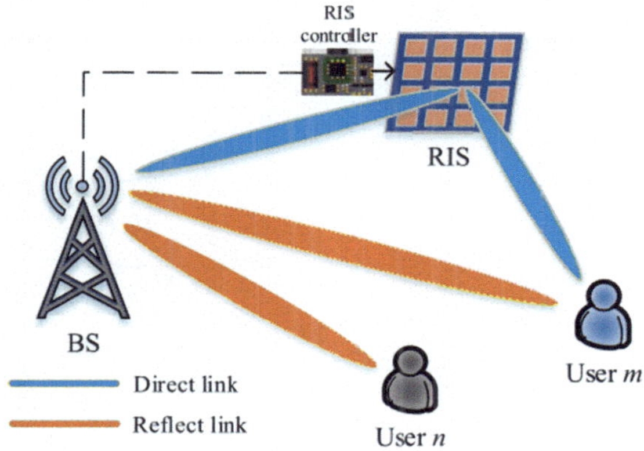


Fig. 1.2 RIS employed to support the communication from the BS to the UEs

## 1.2 Cellular-Connected UAVs

Recently, we have seen a substantial rise in the utilization of unmanned aerial vehicles (UAVs) across civilian applications, encompassing activities such as aerial surveillance, photography, traffic control, communication relay, and package delivery. These regulations are projected to generate over \$82 billion for the U.S. economy and establish more than 100,000 additional employees over the next few years.

However, despite these advancements, several technical challenges must be addressed before widespread UAV deployment becomes practical. Ensuring low-latency, high-capacity, and ultra-reliable dual-way wireless connection between UAVs and their associated ground objects is crucial not only for secure operations but also for facilitating mission-specific, high-data rate-demanding payload communications. Presently, current unmanned aircraft systems predominantly rely on simple point-to-point communication over the unlicensed band (e.g., ISM 2.4 GHz), which poses challenges like minimum data rates, insecurity, unreliability, vulnerability to interference, hurdles in legitimate supervising and managing, and constraints on operational limits.

As the proliferation of UAVs and their miscellaneous applications is anticipated to expand in the foreseeable future, the development of new wireless technologies becomes imperative to enable a considerably improved UAV-ground network. A promising approach is the integration of cellular-connected UAVs into current and future wireless networks as new mid-air or flying BS, as demonstrated in Fig. 1.3. This cellular-connected UAV paradigm offers numerous interesting benefits over traditional communications.

Efforts to integrate UAVs with cellular networks have roots dating back to the 2000s, notably through the Global System for Mobile Communications (GSM). Recently, there has been a significant surge in interest from both the academic world (e.g., [16, 17]) and industry. Qualcomm experimented with drone operations over

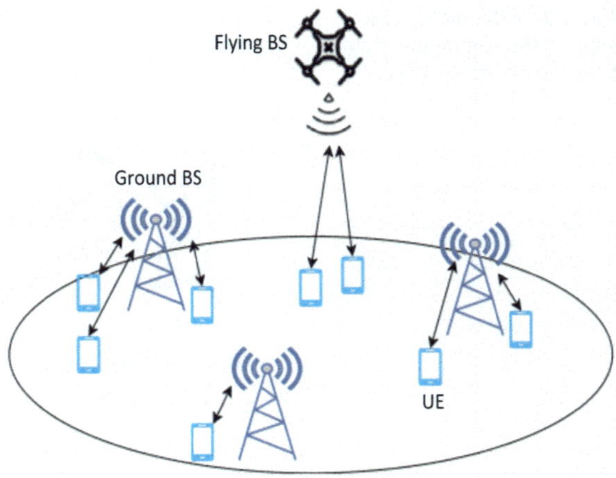


**Fig. 1.3** UAV-assisted cellular communication

commercial LTE networks in September 2016, with an experimental report on LTE unmanned aircraft systems released in May 2017.

In early 2017, the 3rd-Generation Partnership Project (3GPP) approved an advanced work item for studying improved help for aerial vehicles using LTE, leading to the release of several proposals on technical innovations. Notably, UAV-enabled mid-air communication has recently garnered massive research devotion [18]. In this model, committed UAVs function as mid-air BSs, access points, or relays, aiding wireless networks for ground users, referred to as UAV-assisted wireless communication. Previous research has investigated a range of resource allocation strategies for next-generation networks [19–26].

The two concepts of cellular-connected UAVs and UAV-assisted terrestrial communication exhibit both resemblances (e.g., interference and ground-UAV channel characteristics) and essential distinctions. Specifically, the roles of UAVs in such models are inverted: serving as BSs, APs, or relays in UAV-assisted wireless communication versus acting as cellular UE in cellular-enabled UAV communication. The main objective of this book is to provide an innovative and state-of-the-art summary of the promising tools of cellular-connected UAV communication.

The distinctive transmission and spectrum requirements of such wireless communications systems are to be discussed first, followed by innovative design factors and major promising technologies. This aims to enable a future generation of heterogeneous wireless networks to coexist with terrestrial and aerial users in 3D space. Numerical results will support our discussions, and we will conclude by outlining promising directions for further research.

**Table 1.1** Related works in RIS-assisted UAV communications

| References | Scenario and objectives | Optimization targets |
|---|---|---|
| [27] | Maximize system sum rate through joint UAV trajectory, RIS scheduling, and resource allocation. Minimize UAV energy consumption with trajectory, RIS phase shift, power allocation, and decoding order optimization | System sum rate, UAV energy consumption |
| [28] | Achieve secure UAV-ground communication using terrestrial RIS. Optimize UAV trajectory, RIS beamforming, and transmitter power for maximum average worst-case secrecy rate | Average worst-case secrecy rate |
| [29] | Enhance terahertz (THz) communication with UAV and RIS assistance. Optimize UAV trajectory, RIS phase shifts, THz sub-band allocation, and power control for maximum-minimum average rate among all UEs | Minimum average rate among UEs |
| [30] | Compare UAV/RIS configurations (airborne UAV, terrestrial RIS, aerial RIS). Favor aerial RIS for higher average data rate and Line-of-Sight (LoS) probability. Analyze three modes: UAV relaying, RIS reflecting, and combined relaying and reflecting | Average data rate, LoS probability, Ergodic capacity, Energy efficiency |
| [31] | Analyze the performance of aerial RIS system in relaying, reflecting, and combined modes. Develop optimal mode selection based on maximizing ergodic capacity and energy efficiency | Ergodic capacity, energy efficiency |

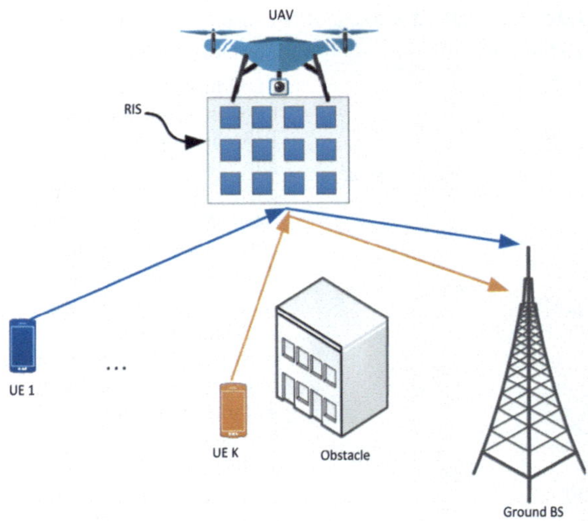


**Fig. 1.4** RIS-assisted UAV communication

## 1.3 Fundamentals of RIS-Assisted UAV Communications

Recently, researchers have been investigating the integration of RISs with UAVs to enhance network performance, as shown in Fig. 1.4. The advantages of combining RISs and UAVs encompass, among other things, improving air-to-ground performance through line-of-sight (LoS) connectivity facilitated by the altitude of the UAVs and any RIS mounted on them; overcoming blockages when users are obstructed from base station signals by obstacles like buildings; and the potential to optimize the deployment of UAVs and RISs for enhanced communication for a larger user base.

An RIS panel can be employed to support a UAV-enabled orthogonal frequency division multiple access (OFDMA) wireless network. The objective of this deployment is to maximize the system sum rate through a joint design of the UAV's trajectory, RIS scheduling, and resource allocation. Moreover, a scenario involving an RIS mounted on a high-rise building to aid a UAV in providing wireless service can be considered. However, the problem of minimizing the UAV's energy consumption, involving RIS phase shift, trajectory design, decoding order, and power allocation, needs to be addressed [27].

A conventional RIS panel deployed on buildings is utilized to establish secure communication between a UAV and a ground user in the presence of an eavesdropper. The aim is to maximize the average worst-case rate concerning the UAV's trajectory, RIS beamforming, and legitimate transmitters' power, employing the alternating optimization technique [28]. A UAV deployed to support terahertz (THz) communication with the assistance of an RIS [29]. There is a need for joint optimization of the RIS phase shifts, UAV's trajectory, power control, and THz sub-band allocation, to maximize the minimum sum capacity of all UEs.

The two possible UAV/RIS settings are flying UAVs, conventional RIS (e.g., RIS mounted on a building facade), and flying RIS (e.g., RIS carried by a flying UAV). Researchers favor flying RISs over conventional RISs in terms of sum capacity and Line-of-Sight (LoS) connection achievable for the downlink between the RIS and UEs [30]. The former provides 360° full-angle reflections, stronger channels, and reduced signal power loss due to fewer reflections [31]. The comparison of related work is summarized in Table 1.1.

The performance of a flying RIS system is analyzed using the following modes: (i) only the UAV performs relaying; (ii) only the RIS performs reflecting; and (iii) the UAV performs relaying and the RIS reflects to receivers. An analytic criterion is derived to optimally select different modes, maximizing the sum capacity and energy efficiency of a wireless system.

## 1.4 Machine Learning for RIS-Assisted UAV Communications

In recent years, the global demand for data traffic has surged dramatically, experiencing a remarkable increase of around 1000 times. This surge is attributed to the evolution of modern wireless communication systems that are anticipated to accommodate the escalating requirements of data traffic in future networks. Prior research has explored various resource allocation strategies [32–43, 45–49, 51–53]. Concurrently, the advent of cutting-edge communication applications, such as wearable devices, autonomous networks, the Internet of Things (IoT), and communications involving fully connected UAVs, continually generates substantial data traffic with diverse needs.

Such evolution in communication scenarios underscores an unavoidable necessity for intelligent processing, service, and optimization in forthcoming communication systems. Integrating machine learning (ML), also known as artificial intelligence, into the design, planning, and optimization of future communication networks is a pivotal response. The concept of ML has a rich and successful history, exemplified by the use of neural networks (NN) for intelligent systems, departing from simplistic single-layer designs to influence the state of multiple neurons.

The application of ML has demonstrated its formidable advantages across various fields, including computer vision, robotics, signal processing, and communications, particularly in scenarios where a comprehensive mathematical model for characteristic patterns is challenging to formulate. In such intricate domains, ML emerges as a potent tool that doesn't necessitate exhaustive model specifications. Traditional methods heavily reliant on models and theories reveal shortcomings due to the intricate nature of communication networks. Consequently, research on ML for communication networks, notably in the realm of wireless communications, is undergoing significant expansion, driven by the need for intelligent processing across diverse scenarios. The evolving framework of ML serves as a facilitator for intelligent pro-

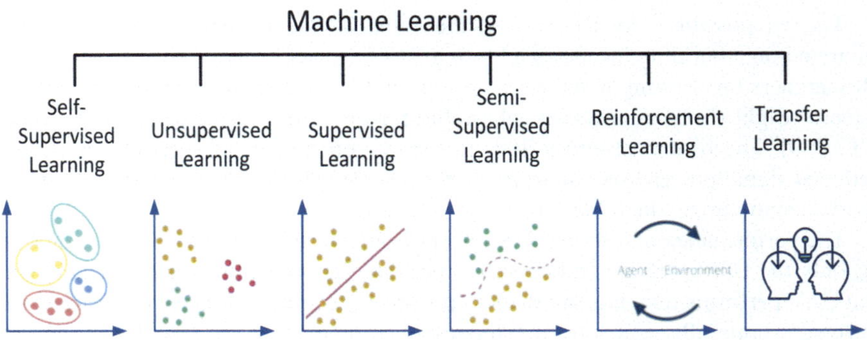


**Fig. 1.5** Primary categories of ML models

cessing in numerous scenarios. Contemporary ML methods offer a broad spectrum of opportunities for developing intelligent communication systems that effectively tackle myriad obstacles related to signal processing, classification, resource allocation, detection, routing protocols, channel estimation, configuring transport protocols, and interpreting user behavior. Different ML schemes find application in radio communication systems, categorized based on the type of supervision required for training the ML models. The pros and cons of various ML models are compared in Table 1.2. The primary categories of ML models are illustrated in Fig. 1.5 and briefly explained below.

### *1.4.1 Unsupervised Learning*

Unsupervised learning in the context of future 6G communication, especially in the domain of RIS and UAV assistance, entails harnessing ML algorithms to unveil inherent patterns and structures within data without explicit labeled guidance. In this setting, RIS is employed to intelligently manipulate wireless signals, and UAVs are deployed to aid and amplify the communication network.

In the realm of unsupervised learning, the model is exposed to unlabeled datasets encompassing various facets of the communication environment, including signal characteristics, interference levels, and network topology. The algorithm independently discerns intrinsic structures and relationships within this data without prior knowledge of specific outcomes.

The integration of unsupervised learning in 6G communication with RIS-UAV assistance empowers the system to unveil concealed patterns and dependencies in the communication environment. For instance, the model may identify optimal configurations of RIS elements or recognize recurring patterns in network traffic without explicit instructions on what to seek.

**Table 1.2** Comparing the pros and cons of various ML models

| Learning model | Cons | Pros |
|---|---|---|
| Supervised learning | 1. Requires a substantial amount of labeled data 2. May not adapt well to dynamic and evolving scenarios | 1. Well suited for tasks with clear objectives and labels 2. Can optimize specific aspects of RIS and UAV assistance 3. Provides precise control over model training |
| Unsupervised learning | 1. A lack of clear objectives may limit specific optimizations 2. Evaluation metrics may be less defined in complex scenarios | 1. Discovers patterns and relationships without explicit labels 2. Useful for exploring unknown aspects of communication 3. Can adapt to changing communication environments |
| Semi-supervised learning | 1. Performance may not match fully supervised approaches 2. Relies on the quality of unlabeled data for effectiveness | 1. Efficiently uses both labeled and unlabeled data 2. Cost-effective compared to fully labeled approaches 3. Adaptable to scenarios with limited labeled data |
| Reinforcement learning | 1. Requires defining a suitable reward structure 2. Computational complexity and training time can be high | 1. Adaptable to sequential decision-making in dynamic environments 2. Can optimize UAV trajectories and RIS configurations 3. Allows for learning from interactions with the environment |
| Self-supervised learning | 1. Designing meaningful self-supervised tasks can be challenging 2. Model performance heavily depends on the chosen self-supervised tasks | 1. Doesn't require explicit labels, useful for scarce labeled data 2. Can leverage large amounts of unlabeled data for training 3. Autonomous discovery of patterns in changing scenarios |
| Transfer learning | 1. Requires careful consideration of domain differences 2. The transfer may not always be beneficial in highly dynamic scenarios | 1. Utilizes knowledge from related tasks for quicker adaptation 2. Reduces the need for extensive labeled data for each task 3. Speeds up adaptation to new communication environments |
| Ensemble learning | 1. Increased computational complexity and resource requirements 2. Coordination and synchronization of ensemble models can be challenging | 1. Enhanced generalization and robustness in complex scenarios 2. Can integrate diverse models for different aspects of 6G communication 3. Reduces overfitting and model variance in changing environments |

By autonomously extracting meaningful information from the data, unsupervised learning contributes to the adaptability and self-optimization of the 6G communication system. It enables the system to dynamically adjust to changing conditions, identify emerging challenges, and unearth novel insights that may elude traditional rule-based approaches. Ultimately, unsupervised learning in this context fosters a more flexible and intelligent communication infrastructure, elevating the overall performance and efficiency of 6G networks with RIS and UAV support.

### *1.4.2 Supervised Learning*

In the context of future 6G communication, particularly concerning RIS and UAV assistance, supervised learning plays a crucial role. This involves the training of ML models using labeled datasets to refine and improve communication systems. Within this framework, RIS is employed to intelligently manipulate wireless signals, while UAVs contribute to supporting and enhancing the communication network.

The supervised learning model undergoes training on datasets featuring diverse input features that depict the communication environment, such as signal strength, interference levels, and network traffic patterns. The corresponding labeled outputs signify the desired communication performance or recommended actions, taking into account factors like signal quality, latency, and reliability.

For supervised learning, the model discerns patterns and correlations between input features and desired outcomes. This attained knowledge is employed in the practical scenarios of UAV-RIs. For example, the optimal configuration of RIS elements is predicted by the model to instruct the UAVs to adjust the optimal position. This accomplished will enhance the signal quality and coverage and reduce the interference.

The core aim of integrating supervised learning in 6G communication with UAV-RIS is to develop the system capability for achieving optimized coverage according to the dynamic environment situations. This accomplishment will result in system throughput performance and ensure reliability, showing the system intelligently operating in 6G communication.

### *1.4.3 Semi-Supervised Learning*

In a semi-supervised learning framework in the context of 6G communication, integrating UAV-RIS, the ML model is trained on datasets that contain both labeled and unlabeled data. The semi-supervised learning model utilizes the guidelines from the labeled datasets to achieve the desired output for such events with unlabeled datasets. This approach gives the model the advantage of the structured guidance given by labeled data while discerning inherent patterns within the unlabeled data.

Semi-supervised learning plays a crucial role in scenarios where obtaining fully labeled datasets is difficult or cost-prohibitive in the context of 6G communication. For instance, the model undergoes partial training, allowing it to glean guidance from labeled data, which aids its performance in analogous scenarios where labeled data is absent. Partial learning enables the model to effectively adapt to various situations within the network, thereby addressing the complexities of UAV-RIS in 6G communication. Therefore, the adaptability in learning equips the model to leverage both labeled and unlabeled data, ensuring optimal communication benefits.

### *1.4.4 Reinforcement Learning*

In reinforcement learning, the ML model engages with the communication environment, receiving feedback to learn and make decisions. This feedback takes the form of rewards for favorable actions and penalties for unfavorable ones. The overarching goal is to attain an optimal policy for accruing rewards over time. Concerning the integration of UAV-RIS in 6G, the model predicts the optimal UAV positioning and configuration of reflection elements. Continuously, it adjusts its actions based on acquired feedback to achieve the overarching objective of enhancing the overall system throughput performance. The reinforcement learning model explores different actions, observes outcomes, and adjusts strategy to maximize the overall communication system. Reinforcement learning is particularly helpful in the dynamic and intricate environment of 6G communication, where conditions swiftly change. Using trial and error, the model uncovers effective strategies for adapting to various challenges. Such adaptive capability is pivotal for optimizing communication systems with the aid of RIS and UAVs, ensuring efficient and reliable performance in evolving communication scenarios.

### *1.4.5 Self-Supervised Learning*

Self-supervised learning performs training of ML models without relying on explicit external labels. Alternatively, the model generates its own labels from the available data. In self-supervised learning, the model is presented with tasks derived from the inherent architecture of the dataset. For 6G communication with RIS-UAV, such tasks include predicting missing components of the communication environment, generating contextually relevant configurations for RIS elements, or predicting the impact of UAV movements on signal quality.

For example, the model might undergo training to predict the next state of the communication environment based on the current state, involving forecasting changes in signal strength or interference levels using historical data. The model essentially learns to create its training signals by comprehending the relationships and patterns present in the data. Self-supervised learning proves advantageous in scenarios where obtaining labeled data is challenging or expensive. Self-supervised learning emerges as the optimal model for tackling the complexities inherent in RIS-UAV communication within 6G networks. This approach allows the model to autonomously learn from the available data, excelling in diverse situations. It inherently discovers the optimal strategies to enhance system performance, particularly in scenarios where explicit labels are sparse.

### *1.4.6 Transfer Learning*

Transfer learning in the context of advancing 6G communication with RIS and UAV assistance involves harnessing knowledge acquired from one task to enhance the performance of a related yet distinct task. In transfer learning, the model grasps the patterns or data of the communication it's trained on. This learning is then leveraged to excel in various tasks. Such knowledge transfer proves highly effective, especially when we contemplate scenarios integrating RIS-UAV in 6G communication.

For instance, the model is pre-trained on a dataset related to general communication patterns, encompassing factors like signal strength, network topology, and interference. Subsequently, the pre-trained model is fine-tuned on a smaller dataset specific to the 6G communication scenario with RIS and UAVs. Transfer of knowledge enables the model to adapt more swiftly to the specifics of the target task, potentially requiring less labeled data for effective training of the ML model.

Transfer learning proves advantageous, especially in scenarios where amassing a large labeled dataset for the target task is impractical and/or costly. It facilitates the reuse of knowledge and patterns learned from related tasks, expediting and streamlining adaptation to the unique challenges posed by 6G communication with RIS and/or UAV. The ability to transfer knowledge between ML models enhances the capacity to optimize communication performance in dynamic and evolving wireless communication environments.

### *1.4.7 Ensemble Learning*

Ensemble learning in the context of advancing 6G communication with RIS and UAV assistance involves the summation of multiple ML models to enhance overall system performance. Ensemble learning entails training several diverse models and combining their predictions to achieve a more robust and accurate outcome than any individual ML model. In the domain of 6G systems, these models are employed to tackle multiple tasks aimed at improving system efficiency. They strive to achieve the optimal configuration of RIS and UAV trajectory, thereby enhancing system effectiveness and reliability.

As an illustration, the ensemble of models encompasses various specialized models. For instance, one model excels at determining the optimal configuration of RIS elements, while another specializes in UAV trajectory planning and positioning. Additionally, there's a model focused on solving network congestion optimally. By combining these diverse models, a robust and reliable model is created, enhancing the reliability and effectiveness of 6G communication systems. Hence, ensemble learning models possess the capability to address challenging and complex situations that individual models may struggle with due to limitations. By integrating protection from different models, ensemble learning minimizes errors, rendering the system more reliable and effective in resolving challenges encountered in 6G communication.

## 1.5 Conclusions

In this chapter, we delve into the study that delineates crucial technologies and research directions relevant to RIS-assisted UAV communication. We present detailed insights into RIS communications and standards, encompassing the transition from physics to wireless communications and the pivotal role of RIS, along with RIS-based beamforming. Additionally, we furnish specifics on how RIS is employed to bolster communication from the BS to the UE for cellular-connected UAVs. Furthermore, we discuss the fundamentals of RIS-assisted UAV communications. Moreover, we elaborate on machine learning models and their application in RIS-Assisted UAV communications. This study underscores the benefits of integrating UAV-RIS into complicated scenarios and emphasizes the significance of employing artificial intelligence technologies for RIS-UAV communication to enhance network throughput performance for 6G communication.

## References

1. Letaief K, Chen W, Shi Y, Zhang J, Zhang Y (2019) The roadmap to 6G: AI empowered wireless networks. IEEE Commun Mag 57:84–90
2. Saad W, Bennis M, Chen M (2020) A vision of 6G wireless systems: applications, trends, technologies, and open research problems. IEEE Netw 34:134–142
3. Wu Q, Zhang R (2020) Towards smart and reconfigurable environment: intelligent reflecting surface aided wireless network. IEEE Commun Mag 58:106–112
4. Cheng Y, Li K, Liu Y, Teh K, Poor HV (2021) Downlink and uplink intelligent reflecting surface aided networks: NOMA and OMA. IEEE Trans Wirel Commun 20:3988–4000
5. Hou T, Liu Y, Song Z, Sun X, Chen Y, Hanzo L (2022) MIMO assisted networks relying on intelligent reflective surfaces: a stochastic geometry based analysis. IEEE Trans Veh Technol 71:571–582
6. Smith D, Yurduseven O, Mancera L, Bowen P, Kundtz N (2017) Analysis of a waveguide-fed metasurface antenna. Phys Rev Appl 8:054048
7. Feng Y, Zhu B, Chen K, Zhao J, Jiang T (2015) Dynamically controlling electromagnetic wave with tunable metasurfaces. In: 2015 international symposium on antennas and propagation (ISAP), pp 1–4

8. Cheng Q (2020) Space-time coding metasurface for wireless communication. In: 2020 IEEE Asia-pacific microwave conference (APMC), pp 3–3
9. Fong B, Colburn J, Ottusch J, Visher J, Sievenpiper D (2010) Scalar and tensor holographic artificial impedance surfaces. IEEE Trans Antennas Propag 58:3212–3221
10. Danufane F, Renzo M, Rosny J, Tretyakov S (2021) On the path-loss of reconfigurable intelligent surfaces: an approach based on green's theorem applied to vector fields. IEEE Trans Commun 69:5573–5592
11. Wu Q, Zhang R (2020) Beamforming optimization for wireless network aided by intelligent reflecting surface with discrete phase shifts. IEEE Trans Commun 68:1838–1851
12. Wu Q, Zhang R (2019) Intelligent reflecting surface enhanced wireless network via joint active and passive beamforming. IEEE Trans Wirel Commun 18:5394–5409
13. Han Y, Tang W, Jin S, Wen C, Ma X (2019) Large intelligent surface-assisted wireless communication exploiting statistical CSI. IEEE Trans Veh Technol 68:8238–8242
14. Huang C, Alexandropoulos G, Zappone A, Debbah M, Yuen C (2018) Energy efficient multi-user MISO communication using low resolution large intelligent surfaces. In: 2018 IEEE Globecom workshops (GC Wkshps), pp 1–6
15. Basar E, Di Renzo M, De Rosny J, Debbah M, Alouini M, Zhang R (2019) Wireless communications through reconfigurable intelligent surfaces. IEEE Access 7:116753–116773
16. Van Der Bergh B, Chiumento A, Pollin S (2016) LTE in the sky: trading off propagation benefits with interference costs for aerial nodes. IEEE Commun Mag 54:44–50
17. Lin X, Yajnanarayana V, Muruganathan S, Gao S, Asplund H, Maattanen H, Bergstrom M, Euler S, Wang Y (2018) The sky is not the limit: LTE for unmanned aerial vehicles. IEEE Commun Mag 56:204–210
18. Zeng Y, Zhang R, Lim T (2016) Wireless communications with unmanned aerial vehicles: opportunities and challenges. IEEE Commun Mag 54:36–42
19. Ishtiaq I, Narmeen R, Aftab MW, Alkhrijah Y, Alawad MA, Kaushik A, Random access optimization with generative adversarial networks in industrial IoT using deep deterministic policy gradient approach. IEEE Open J Commun Soc. https://doi.org/10.1109/OJCOMS.2024.3506782.
20. Ishtiaq I, Narmeen R, Alawad MA, Alkhrijah Y, Ho P-H (2024) Integrating visual geometry and mask region CNN for enhanced UAV detection and identification. In: 2024 IEEE 100th vehicular technology conference (VTC2024-Fall). Washington, DC, USA, pp 1–6. https://doi.org/10.1109/VTC2024-Fall63153.2024.10757938
21. Narmeen R, Mach P, Becvar Z, Ishtiaq I (2024) Joint exit selection and offloading decision for applications based on deep neural networks. IEEE Internet Things J 11(23):38098–38112. https://doi.org/10.1109/JIOT.2024.3444898.
22. Ishtiaq I et al (2024) Optimizing cell association and stability in integrated aerial-to-ground next-generation consumer wireless networks. IEEE Trans Consum Electron 70(3):6262–6276. https://doi.org/10.1109/TCE.2024.3416432
23. Ishtiaq I, Becvar Z, Mach P, Gesbert D (2024) Coordinated machine learning for energy efficient D2D communication. IEEE Wirel Commun Lett 13(5):1493–1497. https://doi.org/10.1109/LWC.2024.3377444
24. Ishtiaq I et al (2024) Machine-learning-based optimal cooperating node selection for internet of underwater things. IEEE Internet Things J 11(12):22471–22482. https://doi.org/10.1109/JIOT.2024.3381834.
25. Ishtiaq I, Becvar Z, Mach P (2024) Coordinated machine learning for channel reuse and transmission power allocation for D2D communication. In: IEEE global communications conference (Globecom)
26. Ishtiaq I, Narmeen R et al (2024) DRL-based resource management for task-centered semantic communication. In: IEEE international symposium on personal, indoor and mobile radio communications (PIMRC workshop)
27. Liu X, Liu Y, Chen Y (2021) Machine learning empowered trajectory and passive beamforming design in UAV-RIS wireless networks. IEEE J Sel Areas Commun 39:2042–2055

28. Li S, Duo B, Renzo M, Tao M, Yuan X (2021) Robust secure UAV communications with the aid of reconfigurable intelligent surfaces. IEEE Trans Wirel Commun 20:6402–6417
29. Pan Y, Wang K, Pan C, Zhu H, Wang J (2021) UAV-assisted and intelligent reflecting surfaces-supported terahertz communications. IEEE Wirel Commun Lett 10:1256–1260
30. Zhang Q, Saad W, Bennis M (2019) Reflections in the sky: millimeter wave communication with UAV-carried intelligent reflectors. In: 2019 IEEE global communications conference (GLOBECOM), pp 1–6
31. Lu H, Zeng Y, Jin S, Zhang R (2021) Aerial intelligent reflecting surface: joint placement and passive beamforming design with 3D beam flattening. IEEE Trans Wirel Commun 20:4128–4143
32. Ahmad I, Kaleem Z, Plachy J, Tabassum H, Becvar Z, Saad W (2024) Guest editorial: flexible reconfigurable intelligent surfaces for integrated air-to-ground internet of things: architecture, recent advances, and future applications. IEEE Internet Things Mag 7(2):12–13
33. Ahmad I et al (2023) Co-channel interference management for heterogeneous networks using deep learning approach. Information 14(139)
34. Ahmad I, Narmeen R, Becvar Z, Guvenc I (2022) Machine learning-based beamforming for unmanned aerial vehicles equipped with reconfigurable intelligent surfaces. IEEE Wirel Commun Mag 29(4):32–38
35. Mughal UA, Ahmad I, Pawase CJ, Chang KH (2022) UAVs path planning by particle swarm optimization based on visual-SLAM algorithm. In: Intelligent unmanned air vehicles communications for public safety networks. Springer
36. Narmeen R, Ahmad I, Costa DBD, Muhaidat S (2021) Shortest propagation delay-based relay selection for underwater acoustic sensor networks. IEEE Access 37923–37935
37. Ahmad I, Jang JY, Chang KH (2020) QoS priority-based mobile personal cell deployment with load balancing for interference reduction between users on coexisting public safety and railway LTE networks. Electronics 9(12)
38. Guo J, Ahmad I, Chang KH (2020) Classification, positioning, and tracking of drones by HMM using acoustic circular microphone array beamforming. J Wirel Commun Netw
39. Ahmad U, Ahmad I, Xiao J, Chang KH (2020) Cooperative resource management for C-V2I communications in a dense urban environment. Veh Commun
40. Teng H, Ahmad I, Msm A, Chang K (2020) 3D optimal surveillance trajectory planning for multiple UAVs by using particle swarm optimization with surveillance area priority. IEEE Access 8:86316–86327. https://doi.org/10.1109/ACCESS.2020.2992217
41. Chang KH, Ahmad I (2020) Mission-critical user priority-based random access method and apparatus for collision resolution in coexisting PS-LTE and LTE-M network. US Patent, 10645734
42. Shi L, Ahmad I, Chang KH (2020) Service group based FOFDM-IDMA platform to support massive connectivity and low latency simultaneously in the uplink IoT environment. Wirel Commun Mob Comput 2020:6. Article ID 8517372. https://doi.org/10.1155/2020/8517372
43. Ahmad I, Chang KH (2020) Mission-critical user priority-based cooperative resource allocation schemes for multi-layer next-generation public safety networks. Phys Commun 38. https://doi.org/10.1016/j.phycom.2019.100926
44. He Y, Chen W, Ahmad I, Shi L, Chang KH (2019) Compressive sensing based random access for machine type communications considering tradeoff between link performance and latency. EURASIP J Wirel Commun Netw. https://doi.org/10.1186/s13638-019-1510-5
45. Ahmad I, Chang KH (2019) Mission critical user priority-based random-access scheme for collision resolution for coexisting PS-LTE and LTE-M networks. IEEE Access 7:115505–115517
46. Ahmad I, Chang KH (2019) Effective SNR mapping and link adaptation strategy for next-generation underwater acoustic communications networks: a cross-layer approach. IEEE Access 7:44150–44164
47. Ahmad I, Chen W, Chang KH (2017) LTE-railway user priority-based cooperative resource allocations schemes for coexisting public safety and railway networks. IEEE Access 5:7958–8000

48. Ahmad I, Chang KH (2019) Downlink power allocation strategy for next-generation underwater acoustic communications networks. Electronics
49. Chen W, Ahmad I, Chang KH (2017) Co-channel interference management using eICIC/FeICIC with coordinated scheduling for the coexistence of PS-LTE and LTE-R networks. EURASIP J Wirel Commun Netw 2017(34):1–14
50. Ahmad I, Nguyen LD, Ha DB (2019) Quality-of-service aware game theory-based uplink power control for 5G heterogeneous networks. Mob Netw Appl 24(2):556–563
51. Shi L, Ahmad I, He Y, Chang KH (2018) Hidden Markov model-based drone sound recognition using MFCC technique in practical noisy environments. J Commun Netw 20(5):509–518
52. He Y, Ahmad I, Shi L, Chang KH (2019) SVM-based drone sound recognition using the combination of HLA and WPT techniques in practical noisy environment. KSII Trans Internet Inf Syst
53. Narmeen R, Ahmad I (2023) Co-channel interference evaluation for coexisting public-safety networks and 5G commercial network. J Smart Internet Things 2:122–132

# Chapter 2
# Securing Air-To-Ground Internet of Things with Artificially Intelligent Reconfigurable Surfaces

**Xin Yuan, Shuyan Hu, Wei Ni, Xin Wang, and Abbas Jamalipour**

## 2.1 Introduction

Reconfigurable Intelligent Surfaces (RISs) and Unmanned Aerial Vehicles (UAVs) hold considerable promise in bolstering the security of Internet-of-Things (IoT) networks. RISs, functioning as intelligent reflectors, passively enhance wireless coverage, while UAVs provide dynamic and adaptable platforms for IoT communication, sensing, and monitoring. This chapter places particular emphasis on the application

X. Yuan
Data61, Commonwealth Scientific and Industrial Research Organization (CSIRO), Sydney, NSW 2122, Australia
e-mail: xin.yuan@csiro.au

X. Yuan · W. Ni
School of Computer Science and Engineering, The University of New South Wales, Sydney, NSW 2052, Australia
e-mail: wei.ni@ieee.org

S. Hu (✉)
College of Electronics and Information Engineering and the Shanghai Institute of Intelligent Science and Technology, Tongji University, Shanghai 201804, China
e-mail: syhu@tongji.edu.cn

W. Ni
School of Engineering, Edith Cowan University, Perth, WA 6027, Australia

X. Wang
Key Laboratory of EMW Information (MoE), College of Future Information Technology, Fudan University, Shanghai 200433, China
e-mail: xwang11@fudan.edu.cn

A. Jamalipour
School of Electrical and Information Engineering, The University of Sydney, Sydney, NSW, Australia
e-mail: a.jamalipour@ieee.org

© The Author(s), under exclusive license to Springer Nature Singapore Pte Ltd. 2026 
I. Ahmad et al. (eds.), *Reconfigurable Intelligent Surfaces for Unmanned Aerial Vehicle Communication*, Unmanned System Technologies,
https://doi.org/10.1007/978-981-96-9662-8_2

of RISs in anti-jamming UAV communication and radio surveillance, addressing challenges traditionally resistant to conventional optimization tools due to their non-convex nature.

To tackle these complexities, novel artificial intelligence (AI) tools, specifically deep reinforcement learning (DRL), are introduced for UAV and RIS design. Leveraging DRL enables UAVs to autonomously learn optimal trajectories and RIS configurations, enabling effective jamming signal mitigation and maximizing communication rates based on received data rates. Moreover, DRL allows UAVs to optimize eavesdropping rates by observing and adapting to the transmit rates of suspicious transmitters during radio surveillance. Moreover, the use of DRL eliminates the need for explicit knowledge of channel state information, enabling UAVs to learn and adapt through trial and error.

### 2.1.1 Air-To-Ground Internet of Things

UAVs are becoming increasingly popular due to their exceptional maneuverability, swift deployment, and extensive coverage capabilities. They are employed in diverse applications such as nature preservation, crisis recovery, and temporary communications [1, 2]. UAVs are installed with advanced onboard cameras and sensing modules like LiDAR and radar, enabling them to track and monitor moving objects effectively, as depicted in Fig. 2.1. These UAV monitoring activities include crime surveillance and precaution, and firefighting.

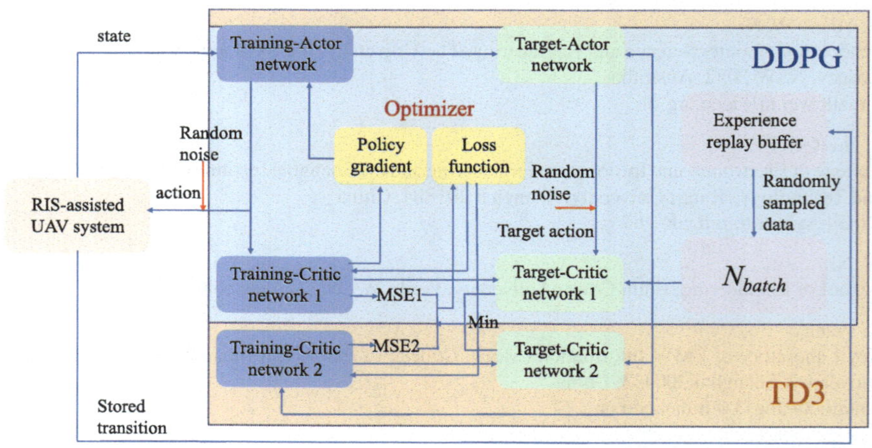


**Fig. 2.1** The architecture of the DDPG- or TD3-based framework, where the actor network is composed of an actor and a target actor, and the critic network of TD3 contains two critics and two target critics (the critic network of DDPG comprises one critic and one target critic). The experience replay buffer provides batches of samples of state transitions to train and update the networks

The utilization of UAVs has seen a rise in conjunction with the expansion of the Internet of Things (IoTs) [3]. One notable example is equipping UAVs with sensors and cameras for monitoring and gathering data on air quality, traffic patterns, and other environmental factors relevant to IoT applications [4]. The collected data by UAVs need to be transmitted back to a control center or ground base station (BS) for analysis or for triggering automated responses, e.g., adjusting traffic lights to alleviate congestion [5]. Some example applications of UAVs are as follows:

- *Safety and security*: Surveillance UAVs can be deployed for various public safety purposes, including highway patrol to detect speeding, nocturnal surveillance, tracking suspicious individuals, and recording incidents of misbehavior.
- *Environment resilience and conservation*: UAVs are invaluable in large-scale disaster or accident areas where manual inspection is challenging. They can aid in wildlife conservation efforts by closely observing animals and recording their natural behaviors.
- *Crisis rescue and recovery*: In the aftermath of earthquakes, deluge, storms, and mudslides, UAVs can be crucial in searching for trapped individuals, assisting rescue teams in navigation, establishing emergency communication networks, and delivering essential supplies. Another promising application is using UAVs for fire fighting in high-rise buildings and forests inaccessible to human firefighters. These UAVs can gather real-time data near the fire site, analyze the situation, and dispense fire-extinguishing chemicals.

### *2.1.2 Reconfigurable Intelligent Surfaces for UAV Communications*

RISs are anticipated to have widespread deployment in the future, primarily for enhancing wireless or cellular communications, as evidenced by their promising performance in studies, such as [6]. RISs can be deployed on building facades and roadside billboards. They passively augment the radio propagation environment to improve communication link conditions and increase data rates. Given the potential of RISs, it is prudent to consider RISs in designing UAV communication systems for developing robust and future-proof technical solutions; please see [6] and the references therein. For instance, a downlink wireless communication system with distributed RISs was studied in [6]. The sum rate was maximized using a penalty-based method and block coordinate descent (BCD) to optimize the UAV movement and power allocation, RIS phase shifts, and decoding order jointly.

UAV-borne RISs have been utilized to facilitate communications among ground nodes. Compared to traditional RISs, UAV-borne RISs offer advantages such as dynamic mobility, a higher likelihood of LoS links from the transmitter to the receiver, and panoramic/full-angle reflection capabilities. However, deploying UAV-borne RISs presents new challenges, including functionality, sustainability, and maneuverability of the UAV platform. Extra flight command units and safety precautions are demanded to operate the UAV-borne RIS. The influence of drift or jittering of the aerial framework on the design of the UAV-borne RIS must be addressed, along with the challenges associated with channel estimation for communication assisted by UAV-borne RISs. These aspects warrant further investigation to advance the understanding and implementation of UAV-borne RIS systems.

The practicality and advantages of RIS-assisted UAV communications were experimentally corroborated by computer simulations in [6] and the references therein. An RIS is typically controlled via a Field Programmable Gate Array (FPGA). The execution delays of FPGA units are in nanoseconds. Therefore, FPGA-controlled RISs can achieve a phase shift change with a delay of microseconds. Consider a UAV's typical speed of about 30 m/s. The reconfiguration of an RIS is fast enough to cater to the time-varying UAV communication channels. It is feasible to introduce RIS into UAV communications.

RISs and UAVs have the potential to play a significant role in enhancing cybersecurity in IoT networks. RISs can be deployed as intelligent reflectors to enhance wireless communication and network coverage. By intelligently manipulating the propagation environment, RISs can mitigate signal interference, improve signal quality, and enhance the overall reliability and security of IoT communications. This capability can be particularly valuable in IoT networks, where many interconnected devices make them vulnerable to potential attacks and security breaches. On the other hand, UAVs can act as mobile sensing and monitoring platforms in IoT networks. They can be equipped with various sensors, cameras, and communication modules to detect and respond to cybersecurity threats. UAVs can actively patrol IoT network areas, identify anomalies, and collect real-time data on potential security breaches. Additionally, UAVs can assist in rapidly deploying security measures in response to emerging threats, such as deploying RISs to enhance network security and resilience.

The combination of RISs and UAVs in IoT networks offers several advantages for cybersecurity. Firstly, RISs provide enhanced signal coverage and quality, reducing the vulnerability of IoT devices to unauthorized access or malicious interference. Secondly, UAVs can serve as agile and adaptable security agents, capable of identifying and responding to security threats in real time. They can conduct surveillance, detect intrusions, and deploy countermeasures to safeguard IoT network integrity. Lastly, the mobility and flexibility of UAVs enable them to address security challenges in diverse and dynamic IoT environments, including remote or hard-to-reach locations. Integrating RISs and UAVs in IoT networks holds great potential for strengthening cybersecurity. RISs enhance wireless communication and network coverage, while UAVs provide mobile sensing and monitoring capabilities. Together, they can improve threat detection, response time, and overall network security, contributing to protecting IoT devices and data from cyberthreats.

## 2.2 RIS for UAV Security: Jamming Rejection and Radio Surveillance

Jamming and radio surveillance are two critical aspects of UAV and air-to-ground network systems. UAVs equipped with communication capabilities play a significant role in various applications, including surveillance, reconnaissance, disaster management, and communication relays. However, these systems are vulnerable to jamming attacks, where malicious entities intentionally disrupt the wireless communication links between UAVs and ground networks.

### *2.2.1 Jamming Rejection*

Jamming can severely affect the physical-layer security of UAV-borne IoT applications [2, 7, 8]. An attacker can use jamming equipment to disrupt the wireless transmissions between the UAVs and the BS, preventing the UAVs from transmitting their sensing data and receiving operation instructions in a timely manner and causing the UAVs to fail their missions [8]. As a matter of fact, jamming has been identified as one of the most critical threats to the massive IoT connections in the context of the upcoming sixth-generation (6G) communication systems [7].

Identifying and eliminating jammers are challenging without specialized equipment, such as radar [3, 9]. Several techniques have been adopted in existing communication systems to alleviate the interference from jammers, such as frequency hopping, power control, relay assistance, and beamforming. Previous research on jamming cancellation for UAV-borne IoT applications has not specifically focused on UAV-borne IoT platforms. In [10], the authors targeted to improve the worst-case long-term data rate for a UAV-assisted wireless sensor network under jamming attacks by collectively configuring the UAV's transmission strategy and 3D flight path. In [11], a reinforcement learning (RL)-based communication strategy was developed between a UAV swarm and a BS to counteract jamming attempts. The communication quality of UAVs was improved by exploring the motion and antenna spatial domain. In [12], the authors proposed a secure UAV communication scheme against smart jammers using a knowledge-based RL approach, which leveraged domain information to reduce the state space and speed up the convergence of the RL algorithm.

### *2.2.2 Radio Surveillance*

Effective radio surveillance is paramount as radio communication remains a dominant method for exchanging messages, including potentially malicious ones [4]. Capturing radio signals enables activities, such as cryptanalysis, traffic analysis, signal signature identification, information integration, and intelligence gathering [2, 13]. UAVs have emerged as a promising choice for surveillance applications due to their high probability of line-of-sight (LoS) radio propagation to/from terrestrial targets. UAV-based radio surveillance enjoys the merits of flexibility and maneuverability. UAVs can access what is beyond stationary terrestrial monitors by dynamically adjusting their positions to avoid barricades and meeting various surveillance requirements, including successful communication rates and covert operations.

Several studies have explored radio and video-based UAV surveillance [2–5, 7, 14]. For instance, the work in [2, 4] aimed to maximize eavesdropping rates for UAV monitors, while the work in [5] focused on minimizing the energy consumption of UAVs for disguised video surveillance.

In applications where UAVs must remain unnoticed by their targets, such as tailing and seizure, animal watching, and digital combat, covert UAV-reliant video monitoring is particularly promising. The advantages of UAV-based covert video surveillance include the following:

- *Flexible and quick dispatch of UAVs*: UAVs can be dispatched rapidly and devise their flight paths dynamically, making them well suited for search and rescue missions, temporary transmissions, and crime surveillance. Unlike stationary ground cameras or automobile-borne cameras, UAVs can freely travel in three-dimensional spaces, substantially diminishing the risk of detection and minimizing the likelihood of targets executing countermeasures.
- *Real-time and precise detection, documentation, and reporting of malpractice*: UAVs equipped with advanced optics and computer vision technologies offer high resolutions, ample storage capacity, and real-time video streaming capabilities. This enables them to detect and archive the malpractice of targets during surveillance operations and immediately send video clips to a central control center. The integrated functionality of real-time video surveillance, recording, and transmission provides a novel and timely solution for evidence collection in criminal forensics.
- *Enhanced automation and software engineering*: Machine learning techniques enable UAVs to possess advanced image processing and manipulation functionalities. The UAVs can recognize targets in real-time and perform various actions, such as measuring target position, predicting trajectory, tracking targets, adjusting camera focal lengths, recording video footage, and transmitting it to the designated location. This integration of machine learning algorithms enhances UAV surveillance's automation and software engineering aspects, allowing for more efficient and intelligent monitoring operations.

On the other hand, RISs can be particularly valuable in radio surveillance. Deployed on building facades and roadside billboards, they passively augment the radio propagation environment without attracting targets' attention, enabling the creation of a covert surveillance channel for monitoring through unlikely means, such as a UAV positioned farther away or obstructed by LoS obstacles.

## 2.2.3 Challenges of RIS-Empowered UAV Security

### 2.2.3.1 Difficulty in CSI Acquisition

One of the challenges is that many RIS-reflecting elements would incur prohibitive overhead and complexity for radio channel acquisition. Let alone the predictive channel estimation or RIS reconfiguration. Consider the real-time requirement of UAV operations. The requirement of instantaneous CSI would pose a significant challenge, especially when a fixed-wing UAV is considered, since a fixed-wing UAV must keep flying to stay afloat (or stalls).

### 2.2.3.2 Limitation of Optimization and Control Techniques

Existing research has explored control-based methods and deep learning techniques for UAV trajectory design, monitoring, and target tracking. Control-reliant approaches, e.g., receding horizon control and dynamic programming, have addressed trajectory planning challenges, while deep learning and DRL have shown promise in target-tracking tasks.

However, several limitations remain unaddressed in the current literature. One reason is that the objective function of successful eavesdropping data rate is a non-smooth, non-convex, piecewise function of the UAV trajectory and RIS phase shifts, which cannot be tackled by the existing optimization techniques, e.g., SDP or SCA. Another reason is that a large number of RIS-reflecting elements incur prohibitive overhead and complexity for radio channel acquisition. Let alone that the CSI of the involved channels is unavailable to the UAV in practice. Furthermore, the online nature of trajectory planning is crucial, especially in the presence of adversaries. Real-time adjustments to the UAV's trajectory are often required, necessitating online planning and optimization. Traditional optimization techniques are not well suited for online trajectory planning [13].

Moreover, the involvement of RISs in UAV communications introduces additional challenges. Specifically, the waypoints of the UAV are nested in the exponents of the array responses of the RIS-reflected channels, making the trajectory optimization intractable for existing convexification tools, such as SDP, Lagrange-dual method, SCA, or manifold optimization (MO). Many RIS-reflecting elements also result in complexity and overhead in radio channel acquisition. Further, the lack of accurate CSI complicates the optimization process. These challenges make it difficult to solve UAV trajectory planning and RIS configuration jointly using existing optimization techniques.

## 2.3 AI for RIS-Empowered UAV Security

A DRL framework can be employed to address the challenges posed by the non-smooth, non-convex objective function and the nested waypoints in the UAV trajectory and RIS configuration problem. Algorithms, such as DDPG and TD3, can be utilized to efficiently learn the optimal UAV trajectory and RIS configuration online [15].

As illustrated in Fig. 2.3, TD3 addresses the Q-value overestimation problem of the DDPG algorithm by incorporating *three improvements* over the classical DDPG model, namely, *clipped double-Q learning*, *target policy smoothing*, and *delayed policy update* [15]. TD3 contains two training-critic and target-critic networks to produce two Q-values. The lesser of the two is used to evaluate the target Q-value in the Bellman error loss function. TD3 perturbs actions produced by the target-actor network (i.e., "target action") with noises. It smooths the corresponding Q-function values to enhance the resistance of the policy against erroneous Q-functions. The training-actor and target-actor networks (i.e., policies) are refreshed less frequently than the training-critic and target-critic networks.

In contrast, the classical DDPG model only contains a training-critic network and a target-critic network, and produces a Q-value. The DDPG model does not add noises toward target actions. Moreover, the DDPG model refreshes its train-actor and target-actor networks and train-critic and target-critic networks simultaneously. As a consequence, TD3 is expected to improve network stability and sample efficiency in continuous control tasks when compared to DDPG.

Some other DRL techniques can be applied to UAV navigation and RIS configuration. These algorithms, including Proximal Policy Optimization (PPO), Trust Region Policy Optimization (TRPO), Model-Based Reinforcement Learning (MBRL), Soft Actor-Critic (SAC), and Reinforcement Learning from Human Feedback (RLHF), offer different advantages for solving complex optimization problems. PPO uses a surrogate objective function to update policies, ensuring stable and efficient learning by balancing exploration and exploitation. TRPO maximizes performance while limiting policy changes during updates to maintain stability. MBRL combines model-free and model-based methods by learning environment dynamics, enabling efficient exploration and exploitation. SAC incorporates entropy regularization to encourage exploration and robust policy learning. RLHF involves human feedback to guide the learning process and improve training in UAV navigation and RIS configuration.

DDPG and its enhanced variation, namely TD3, can be employed to train the UAV for navigation and RIS configuration. The reason is that traditional optimization methods, such as SDP and SCA, are inapplicable because the objective function of the UAV navigation and RIS configuration is non-smooth and non-convex. While AI, more explicitly, DRL, has demonstrated extraordinary capabilities in general to control system operations, the suitability of individual DRL models still needs to be assessed due to their individual strengths and limitations. On the one hand, PPO, TRPO, and SAC are based on stochastic policy gradient, and may not be efficient in complex systems with large state and action spaces. On the other hand, MBRL may suffer from overfitting by exploiting the model too much, and RLHF is expensive as it requires human labeling and feedback [16].

In general, our DDPG/TD3-based approach involves learning to adjust the RIS and control the UAV's trajectory, including heading and acceleration, based on changes in the received data rate of the UAV. The approach eliminates the need for precise CSI or knowledge of the RIS reflecting channels. The reward function depends on the UAV data rate and distance to a predefined destination. The state space contains the UAV's distance to the final location, speed, and data rate. The action space contains the UAV acceleration and RIS phase shifts.

## 2.3.1 AI for RIS-Based Jamming Rejection

### 2.3.1.1 System Model

As depicted in Fig. 2.2, a ground BS offers wireless communication service for a fixed-wing UAV in the presence of a terrestrial jammer. Consider a three-dimensional (3D) Cartesian coordinate system. The BS and the jammer are placed at $\boldsymbol{q}_{\mathrm{B}} = [0, 0, 0]^T$ and $\boldsymbol{q}_{\mathrm{J}} = [x_{\mathrm{J}}, y_{\mathrm{J}}, 0]^T$. An RIS is deployed to facilitate the UAV communication and suppress the jamming signals from the jammer. We suppose that the BS, UAV and jammer all have a single antenna for description convenience (but the DRL architecture can be extended to support multi-antenna BSs and UAVs as part of our future work). Only based on its received data rate does the UAV determine its flight path and the RIS configuration dynamically. The RIS is installed with a smart controller to procure the command from the UAV (via the cellular system) for RIS configuration [17].

**UAV Mobility Model**

The UAV functions for a finite scheduling horizon of $T$ seconds, which is split into $T_w$ time slots indexed by $t$, and $t = 1, \ldots, T_w$. A slot lasts $\delta = T/T_w$, which is short enough that the UAV can be viewed as stationary per slot. The UAV is traveling from a predetermined starting position $\boldsymbol{q}_0 = [x_0, y_0, z_0]^T$ to a predetermined ending position $\boldsymbol{q}_F = [x_F, y_F, z_F]^T$. The UAV's 3D coordinates are $\boldsymbol{q}_t = [x_t, y_t, z_t]^T$, $\forall t$. Let $\boldsymbol{V}_t := [V_{xt}, V_{yt}, V_{zt}]^T$ and $\boldsymbol{A}_t := [A_{xt}, A_{yt}, A_{zt}]^T$ collect the velocity and acceleration of the UAV per slot $t$, respectively. The UAV obeys some mobility constraints [18]:

$$\boldsymbol{q}_{t+1} = \boldsymbol{q}_t + \boldsymbol{V}_t\delta + \frac{1}{2}\boldsymbol{A}_t\delta^2, \ \forall t, \tag{2.1a}$$

$$\boldsymbol{q}_F = \boldsymbol{q}_{T_w}, \tag{2.1b}$$

$$\boldsymbol{V}_{t+1} = \boldsymbol{V}_t + \boldsymbol{A}_t\delta, \ \forall t, \tag{2.1c}$$

$$|A_{i,t}| \leq A_{\max}, i \in \{x, y, z\}, \ \forall t, \tag{2.1d}$$

$$V_{\min} \leq \|\boldsymbol{V}_t\| \leq V_{\max}, \ \forall t, \tag{2.1e}$$

$$V_{zt}/\|\boldsymbol{V}_t\| \leq \sin\vartheta, \ \forall t, \tag{2.1f}$$

where $A_{\max}$ is the UAV's largest acceleration; $V_{\max}$ and $V_{\min}$ are the UAV's maximum and minimum speeds; and $\vartheta$ is the largest UAV pitch angle when ascending or descending.

**RIS Configuration**

The RIS adheres on the outer surface of a building, which is on the $(x, z)$-plane and aligns with the $x$-axis. The RIS has a uniform rectangular array (URA) of $N = N_x N_y$ reflecting elements (units), and a controller to dynamically control the phase shift of each unit. Let $\boldsymbol{\Theta}_t := \mathrm{diag}(e^{j\theta_1^t}, \ldots, e^{j\theta_N^t})$ be the phase shift matrix for the RIS at time slot $t$, where $\theta_n^t = \theta_{(n_x-1)N_y+n_y}^t \in [-\pi, \pi), n = 1, \ldots, N$ is the phase shift of the $n$-th reflecting unit which is located at the $n_y$-th row and the $n_x$-th column of the RIS, and $j = \sqrt{-1}$. The first element of the RIS is at the right bottom corner of the RIS, and its coordinates are $\boldsymbol{q}_{\mathrm{R}} = [x_{\mathrm{R}}, y_{\mathrm{R}}, z_{\mathrm{R}}]^T$.

**Channel Model**

The distances of the BS-UAV link $d_{\mathrm{BU}}^t$, the BS-RIS link $d_{\mathrm{BR}}$, the Jammer-RIS link $d_{\mathrm{JR}}$, the Jammer-UAV link $d_{\mathrm{JU}}^t$, and RIS-UAV link $d_{\mathrm{RU}}^t$ are given by

$$
\begin{aligned}
d_{\mathrm{BU}}^t &= \|\boldsymbol{q}_t\|, \ \forall t, \\
d_{\mathrm{BR}} &= \|\boldsymbol{q}_{\mathrm{R}}\|, \\
d_{\mathrm{JR}} &= \|\boldsymbol{q}_{\mathrm{J}} - \boldsymbol{q}_{\mathrm{R}}\|; \qquad (2.2\mathrm{a}) \\
d_{\mathrm{JU}}^t &= \|\boldsymbol{q}_t - \boldsymbol{q}_{\mathrm{J}}\|, \\
d_{\mathrm{RU}}^t &= \|\boldsymbol{q}_t - \boldsymbol{q}_{\mathrm{R}}\|, \ \forall t. \qquad (2.2\mathrm{b})
\end{aligned}
$$

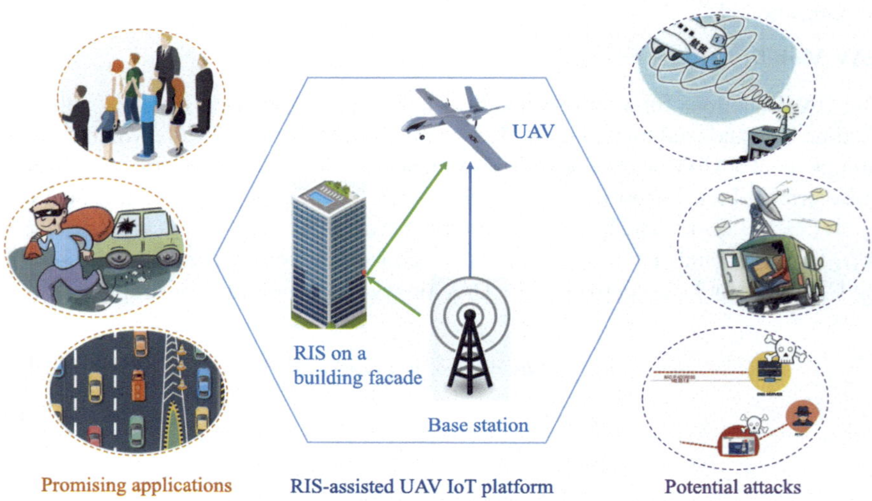


**Fig. 2.2** Schematic of a secure RIS-assisted UAV surveillance network

Consider an LoS channel for the RIS-UAV link (i.e., the R-U link), and Rician fading channels between the BS/jammer and the UAV (i.e., the B-U and J-U links), and between the BS/jammer and the RIS (i.e., the B-R and J-R links). The channel gains of the B-U and J-U links are: $\forall t$,

$$h_{\mathrm{BU}}^{t} = \sqrt{\rho(d_{\mathrm{BU}}^{t})^{-\kappa_1}} \left( \sqrt{\frac{\beta_t}{1+\beta_t}} g_{\mathrm{BU}}^{t} + \sqrt{\frac{1}{1+\beta_t}} \tilde{g}_{\mathrm{BU}}^{t} \right), \tag{2.3a}$$

$$h_{\mathrm{JU}}^{t} = \sqrt{\rho(d_{\mathrm{JU}}^{t})^{-\kappa_1}} \left( \sqrt{\frac{\beta_t}{1+\beta_t}} g_{\mathrm{JU}}^{t} + \sqrt{\frac{1}{1+\beta_t}} \tilde{g}_{\mathrm{JU}}^{t} \right), \tag{2.3b}$$

where $\rho$ stands for the path loss at the reference distance $d_0 = 1$ m with the path loss exponent $\kappa_1 > 2$; $\beta_t$ is the Rician factor of the B-U and J-U links; $g_{\mathrm{BU}}^{t}$ and $g_{\mathrm{JU}}^{t}$ are the deterministic LoS components with $|g_{\mathrm{BU}}^{t}| = 1$ and $|g_{\mathrm{JU}}^{t}| = 1$; and $\tilde{g}_{\mathrm{BU}}^{t}$ and $\tilde{g}_{\mathrm{JU}}^{t}$ stand for stochastic dispersion captured by a zero-mean, unit-variance circularly symmetric complex Gaussian (CSCG) random variable.

The elevation-angle-reliant Rician factor $\beta_t$ is captured by the following exponential function [19]:

$$\beta_t = \xi_1 \exp\left(\xi_2 \arcsin(z_t/d_{\mathrm{BU}}^{t})\right), \ \forall t, \tag{2.4}$$

where $\xi_1$ and $\xi_2$ are two constant coefficients dependent on the environment.

The channel gains of the B-R, J-R, and R-U links, denoted by $\boldsymbol{h}_{\mathrm{BR}} \in \mathbb{C}^{N\times 1}$, $\boldsymbol{h}_{\mathrm{JR}} \in \mathbb{C}^{N\times 1}$, and $\boldsymbol{h}_{\mathrm{RU}}^{t} \in \mathbb{C}^{N\times 1}$, are

$$\boldsymbol{h}_{\mathrm{BR}} = \underbrace{\sqrt{\rho d_{\mathrm{BR}}^{-\kappa_2}}}_{\text{path loss}} \underbrace{\left( \sqrt{\frac{\beta}{1+\beta}} \boldsymbol{h}_{\mathrm{BR}}^{los} + \sqrt{\frac{1}{1+\beta}} \boldsymbol{h}_{\mathrm{BR}}^{nlos} \right)}_{\text{array response \& small-scale fading}}; \tag{2.5a}$$

$$\boldsymbol{h}_{\mathrm{JR}} = \underbrace{\sqrt{\rho d_{\mathrm{JR}}^{-\kappa_2}}}_{\text{path loss}} \underbrace{\left( \sqrt{\frac{\beta}{1+\beta}} \boldsymbol{h}_{\mathrm{JR}}^{los} + \sqrt{\frac{1}{1+\beta}} \boldsymbol{h}_{\mathrm{JR}}^{nlos} \right)}_{\text{array response \& small-scale fading}}; \tag{2.5b}$$

$$\boldsymbol{h}_{\mathrm{RU}}^{t} = \sqrt{\rho(d_{\mathrm{RU}}^{t})^{-2}} \boldsymbol{g}_{\mathrm{RU}}^{t}, \ \forall t. \tag{2.5c}$$

Here, $\beta$ is the Rician factor of the B-R and J-R links (c.f. $\beta_t$), and $\boldsymbol{h}_{\mathrm{BR}}^{los}$ and $\boldsymbol{h}_{\mathrm{JR}}^{los}$ are the LoS components, as given by

$$\boldsymbol{h}_{\mathrm{BR}}^{los} = \left[1, \ldots, e^{-j\frac{2\pi d_x}{\lambda}(N_x-1)\phi_{\mathrm{BR}}^{x}}\right]^T \otimes \left[1, \ldots, e^{-j\frac{2\pi d_y}{\lambda}(N_y-1)\phi_{\mathrm{BR}}^{y}}\right]^T, \tag{2.6a}$$

$$\boldsymbol{h}_{\mathrm{JR}}^{los} = \left[1, \ldots, e^{-j\frac{2\pi d_x}{\lambda}(N_x-1)\phi_{\mathrm{JR}}^{x}}\right]^T \otimes \left[1, \ldots, e^{-j\frac{2\pi d_y}{\lambda}(N_y-1)\phi_{\mathrm{JR}}^{y}}\right]^T, \tag{2.6b}$$

where $d_x$ and $d_y$ are the antenna spacings in the directions of the $x$- and $y$-axes, respectively; $\phi_{\mathrm{BR}}^{x} = x_{\mathrm{R}}/d_{\mathrm{BR}}$ and $\phi_{\mathrm{BR}}^{y} = y_{\mathrm{R}}/d_{\mathrm{BR}}$ are the spatial frequencies corresponding to the angles-of-arrivals (AoAs) from BS to RIS, and $\phi_{\mathrm{JR}}^{x} = (x_{\mathrm{J}} - x_{\mathrm{R}})/d_{\mathrm{JR}}$ and $\phi_{\mathrm{JR}}^{y} = (y_{\mathrm{J}} - y_{\mathrm{R}})/d_{\mathrm{JR}}$ are the spatial frequencies corresponding to the AoAs from the jammer to the RIS along the $x$- and $y$-axes, respectively [20].

In (2.24a) and (2.24b), $\boldsymbol{h}_{\mathrm{BR}}^{nlos} \in \mathbb{C}^{N\times 1}$ and $\boldsymbol{h}_{\mathrm{JR}}^{nlos} \in \mathbb{C}^{N\times 1}$ are the non-LoS (NLoS) components with the variables independently drawn from the zero-mean, unit-variance CSCG distribution. In (2.5c), $\boldsymbol{g}_{\mathrm{RU}}^{t}$ is the array response, as given by

$$\boldsymbol{g}_{\mathrm{RU}}^{t} = \left[1, \ldots, e^{-j\frac{2\pi d_x}{\lambda}(N_x-1)\phi_{\mathrm{RU},x}^{t}}\right]^T \otimes \left[1, \ldots, e^{-j\frac{2\pi d_y}{\lambda}(N_y-1)\phi_{\mathrm{RU},y}^{t}}\right]^T, \ \forall t, \tag{2.7}$$

where $\phi_{\mathrm{RU},x}^{t} = (x_t - x_{\mathrm{R}})/d_{\mathrm{RU}}^{t}$ and $\phi_{\mathrm{RU},y}^{t} = (y_t - y_{\mathrm{R}})/d_{\mathrm{RU}}^{t}$ are the spatial frequencies corresponding to the angles-of-departures (AoDs) from RIS to UAV along the $x$- and $y$-axes, respectively.

It is noteworthy that the UAV does not ask for the CSI involving the RIS and the jammer to produce its flight path and configure the RIS. Instead, the UAV only measures its own received data rate to learn the control policy of its flight path and RIS configuration. This consideration is of practical value since the RIS-reflected channels are difficult and slow to estimate.

#### 2.3.1.2 Problem Formulation

Let $P_t, \forall t$ denote the transmit power of the BS and $P_{\mathrm{J}}$ denote the transmit power of the jammer. The signal-to-interference-plus-noise ratio (SINR) at the UAV at time slot $t$, denoted by $\gamma_{\mathrm{U}}^{t}$, is

$$\gamma_{\mathrm{U}}^{t} = \frac{P_t |h_{\mathrm{BU}}^{t} + (\boldsymbol{h}_{\mathrm{RU}}^{t})^H \boldsymbol{\Theta}_t \boldsymbol{h}_{\mathrm{BR}}|^2}{P_{\mathrm{J}} |h_{\mathrm{JU}}^{t} + (\boldsymbol{h}_{\mathrm{RU}}^{t})^H \boldsymbol{\Theta}_t \boldsymbol{h}_{\mathrm{JR}}|^2 + \sigma_{\mathrm{U}}^2}, \tag{2.8}$$

where $\sigma_{\mathrm{U}}^2$ is the variance of the additive white Gaussian noise (AWGN) at the UAV. The received data rate of the UAV at time slot $t$ is given by

$$R_{\mathrm{U}}^{t} = \log_2(1 + \gamma_{\mathrm{U}}^{t}). \tag{2.9}$$

We aim to maximize the total received data rate of the UAV from the BS over the mission duration of $T_w$ slots. The problem considered is stated as follows:

$$\max_{\{\boldsymbol{q}_t, \boldsymbol{\Theta}_t, \forall t\}} \sum_{t=1}^{T_w} R_{\mathrm{U}}^{t} \tag{2.10a}$$

$$\text{s.t.} \ -\pi \le \theta_n^t < \pi, \ \forall n, t, \tag{2.10b}$$

$$(1a) - (1f).$$

Problem (2.10) is challenging for traditional convex solvers due to several reasons: First of all, the received data rate is a non-convex function of the UAV's flight path $\boldsymbol{q}_t, \forall t$ and the RIS phase shifts $\boldsymbol{\Theta}_t, \forall t$. Second, the UAV flight path waypoints are embedded in the exponents of the R-U link in (2.5c) and (2.28), making the trajectory optimization intractable for existing convex tools, such as successive convex approximation. Another reason is that the large number of RIS reflecting units can cause prohibitive overhead and complexity for radio channel estimation, acquisition, and reconfiguration. To overcome these limitations, the next section describes using DRL to solve (2.10).

#### 2.3.1.3 DRL for Anti-Jamming Communication of UAV-Borne IoT Platforms

The described method in this section aims to solve problem (2.10) by utilizing the DDPG and TD3 models. Our approach involves learning to adjust the RIS and control the UAV's trajectory, including heading and acceleration, based on changes in the received data rate of the UAV. Importantly, our method eliminates the need for precise CSI or knowledge of the RIS-reflecting channels. DDPG and its variations, such as TD3, have been demonstrated to be effective in addressing problems with continuous action spaces [15, 21, 22]. In contrast, traditional DRL methods, such as deep Q learning, can struggle and even diverge when faced with continuous action spaces.

**State, Action, and Reward**

Since the current UAV location only depends on its previous location and speed, the UAV trajectory (i.e., waypoints) is a Markov decision process (MDP). The RIS configuration depends solely on the instantaneous position of the UAV. Therefore, we interpret problem (2.10) as an MDP with its state, action, and reward defined below.

- ***State Space*** $\mathcal{S}$: At time slot $t$, the system state $s_t \in \mathcal{S}$ is made of the relative position of the UAV with regard to its final location, $\boldsymbol{q}_t - \boldsymbol{q}_F$; the velocity of the UAV, $\boldsymbol{V}_t$; and the SINR at the UAV, $\gamma_{\text{U}}^t$, $s_t = \{\boldsymbol{q}_t - \boldsymbol{q}_F, V_t, \gamma_{\text{U}}^t\}$.
- ***Action Space*** $\mathcal{A}$: It gathers all possible actions, i.e., $a_t \in \mathcal{A}$. During the $t$-th time step, the action $a_t$ consists of the reflecting coefficients $\{\theta_n^t\}_{n\in\mathcal{N}}$ and the acceleration of the UAV, $\boldsymbol{A}_t := [A_{xt}, A_{yt}, A_{zt}]^T$, i.e., $a_t = \{\theta_n^t \in [-\pi, \pi), \forall n,$ $A_{it} \in [-A_{\max}, A_{\max}],\ i \in \{x, y, z\}\}$. The UAV acceleration is constrained by (2.1d)–(2.1f). Given the initial location and velocity of the UAV, its future waypoints $\boldsymbol{q}_t$ and velocities $\boldsymbol{V}_t$ are decided by the accelerations, i.e., by (2.1a)–(2.1c).
- ***Reward*** $r_t$: The reward function gives positive returns per time step for implementing action $a_t$:

$$r_t = \underbrace{R_{\text{U}}^t}_{\text{communication}} + \underbrace{\zeta\left(d_F^{t-1} - d_F^t\right)}_{\text{distance to the final location}}, \tag{2.11}$$

where $R_U^t$ is the UAV's received data rate; $d_F^{t-1} = \|\boldsymbol{q}_{t-1} - \boldsymbol{q}_F\|$ and $d_F^t = \|\boldsymbol{q}_t - \boldsymbol{q}_F\|$ are the distances from the UAV to the final location at the $(t-1)$-th and $t$-th time steps, respectively; and $\zeta$ is a tunable parameter during the learning process. The second element on the right-hand side of (2.11) encourages the UAV to fly toward the final location.

- ***Policy***: A projection from the state space, $\mathcal{S}$, to the action space $\mathcal{A}$ is referred to as a policy, $\mu : \mathcal{S} \to \mathcal{A}$, a distribution $\mu(a|s) = \Pr(a_t = a|s_t = s)$ over state $s \in \mathcal{S}$.
- ***Experience***: The experience, defined as $e_t = (s_t, a_t, r_t, s_{t+1})$, is stored in an experience replay memory $\boldsymbol{R}$.

The UAV experiences state $s_t$, performs action $a_t$, receives reward $r_t$, and turns to state $s_{t+1}$. A policy $a_t = \mu(s_t)$ maps state $s_t$ to a possible action. The UAV chooses the policy that maximizes the cumulative reward $R_t = \sum_{n=t}^{N} \gamma^{n-t} r_t$. Here, $\gamma \in (0, 1)$ gives the discount factor. Given $s_t$, $a_t$, and $\mu$, the Q-function evaluates $R_t$ by

$$Q_\mu(s_t, a_t) = \mathbb{E}_\mu[R_t|s_t, a_t]. \tag{2.12}$$

The action-value function, $Q_\mu(s_t, a_t)$, follows the Bellman expectation equation:

$$Q_\mu(s_t, a_t) = \mathbb{E}_{r_t, s_{t+1} \sim \mathcal{E}}\left[r_t + \gamma \mathbb{E}_{a_{t+1} \sim \mu}\left[Q_\mu(s_{t+1}, a_{t+1})\right]\right]. \tag{2.13}$$

Here, $\mathcal{E}$ stands for the environment the UAV experiences.

It is generally challenging to directly use an RL algorithm to solve the continuous-space, finite-horizon MDP and determine the Q-value, $Q(s_t, a_t)$, due to the continuous state and action spaces. A DDPG-based algorithm can be adopted to control the UAV's trajectory and configure the RIS, as delineated in the following subsection.

**Actor-Critic Framework-Based DDPG**

The DDPG-based network uses four DNN approximators, including training-actor and training-critic networks, and target-actor and target-critic networks, as shown in

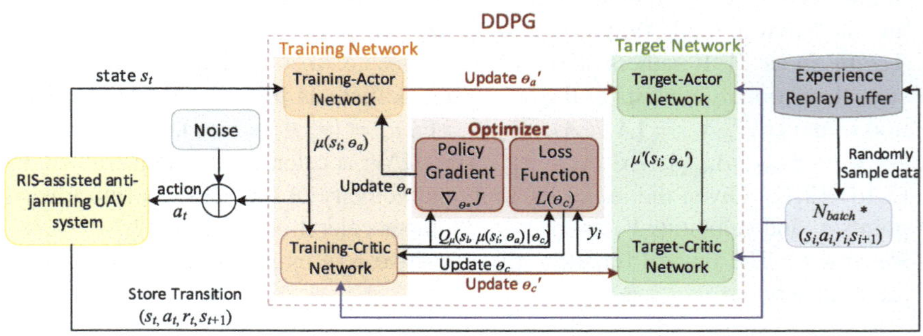


**Fig. 2.3** DDPG-based framework with a training network and a target network, each comprising an actor network and a critic network. The experience replay buffer gives batches of samples of state transitions for training and updating the networks

Fig. 2.3. The training-actor network with parameters $\theta_a$, denoted as $\mu\left(s_t;\theta_a\right)$, gives an approximate policy of the UAV and produces the actions. The training-critic network with parameters $\theta_c$, denoted as $Q_\mu(s_t, a_t; \theta_c)$, estimates the action-value function concerning the actions created in the training-actor network [21]. The target-actor network with parameter $\theta_a'$, represented by $\mu'\left(s_t;\theta_a'\right)$, and the target-critic networks with parameter $\theta_c'$, represented by $Q'_{\mu'}(s_t, a_t; \theta_c')$, generate the target Q-value for training the training-actor and training-critic networks.

The DDPG network uses the deterministic policy gradient (DPG) theorem [21] to refresh $\theta_a$, $\theta_c$, $\theta_a'$, and $\theta_c'$. It produces actions in an actor-critic setting. Additionally, the adoption of a target network (i.e., the target-actor and target-critic networks) helps prevent unstable learning, as opposed to using only a training network (with a training-actor and a training-critic network) [23].

The UAV inputs state $s_t$ into the training-actor network. Using the DPG theorem [21], the network generates the strategy by projecting the state to an action in a deterministic fashion. This network approximates the agent's policy function and selects action $a_t$. Noise is added to $a_t$ to balance between new and known actions, resulting in an output action $a_t = \mu\left(s_t;\theta_a\right) + \mathcal{N}_t$. Herein, $\mathcal{N}_t$ is a random noise process with a normal distribution. The agent is rewarded with $r_t$ and transitions to state $s_{t+1}$. Then, it stores the experience $(s_t, a_t, r_t, st+1)$ in $\boldsymbol{R}$.

The training-critic network evaluates the action-value function $Q_\mu\left(s_t, \mu(s_t;\theta_a);\theta_c\right)$ of the selected action $a_t$. By using a random sample from the replay memory $\boldsymbol{R}$, the network approximates the action-value function as $Q_\mu(s_i, \mu(s_i;\theta_a);\theta_c)$. We take $J(\theta_a)$ as the probability distribution of the parameter $\theta_a$. The training-actor network is adjusted in the direction that improves the strategy the most rapidly, i.e., in the direction of the gradient of $J(\theta_a)$ with respect to (w.r.t.) $\theta_a$ [21]:

$$\nabla_{\theta_a} J(\theta_a) = \mathbb{E}_{s\sim\rho^\mu}\left[\nabla_{\theta_a} Q_\mu(s_t, \mu(s_t;\theta_a);\theta_c)\right] \tag{2.14a}$$

$$= \mathbb{E}_{s\sim\rho^\mu}\left[\nabla_{\theta_a}\mu(s_t;\theta_a)\nabla_a Q_\mu(s_t, a;\theta_c)|_{a=\mu(s_t;\theta_a)}\right], \tag{2.14b}$$

where (2.14b) uses the chain rule; $\rho^\mu$ provides a discounted state distribution of $\mu(s_t;\theta_a)$ [22]; $\nabla_{\theta_a}\mu(s)$ gives the gradient of the training-actor network $\mu(s)$ w.r.t. $\theta_a$; and $\nabla_a Q_\mu(s_t, a;\theta_a)$ provides the gradient of $Q_\mu(s_t, a;\theta_a)$ w.r.t. $a$.

By randomly drawing $N_{batch}$ sampled historical transitions from $\boldsymbol{R}$, the gradient $\nabla_{\theta_a} J(\theta_a)$ is approximated by

$$\nabla_{\theta_a} J(\theta_a) \approx \frac{1}{N_{batch}} \sum_{i=1}^{N_{batch}} \left[\nabla_{\theta_a}\mu(s_i)\nabla_a Q_\mu(s_i, a;\theta_c)|_{a=\mu(s_i)}\right]. \tag{2.15}$$

The training-actor network parameter, i.e., $\theta_a$, is refreshed based on the gradient ascent [24]

$$\theta_a \leftarrow \theta_a + \eta_a \nabla_{\theta_a} J(\theta_a)$$
$$\approx \theta_a + \frac{\eta_a}{N_{batch}} \sum_{i=1}^{N_{batch}} \left[ \nabla_{\theta_a} \mu(s_i) \nabla_a Q_\mu(s_i, a; \theta_c)|_{a=\mu(s_i)} \right]. \tag{2.16}$$

Here, $\eta_a$ specifies the learning rate of the training-actor network.

The training-critic network is updated by minimizing the following loss function:

$$L(\theta_c) = \mathbb{E}_{s_t \sim \rho^\mu, a_t \sim \mu(s_t;\theta_a)} \left[ \left( Q_\mu(s_t, a_t; \theta_c) - y_t \right)^2 \right]. \tag{2.17}$$

Here, $y_t = r_t + \gamma Q'_{\mu'} \left( s_{t+1}, \mu' \left( s_{t+1}; \theta'_a \right); \theta'_c \right)$ is the target Q-value produced by the target network under the transition $(s_t, a_t, r_t, s_{t+1})$. Here, the parameters of the target-actor and target-critic networks, $\theta'_a$ and $\theta'_c$, are the respective decayed copies of $\theta_a$ and $\theta_c$.

With $N_{batch}$ randomly sampled transitions, the loss function, $L(\theta_c)$, is approximately evaluated by

$$L(\theta_c) \approx \frac{1}{N_{batch}} \sum_{i=1}^{N_{batch}} \left[ \left( Q_\mu(s_i, \mu(s_i; \theta_a); \theta_c) - y_i \right)^2 \right], \tag{2.18}$$

where $y_i = r_i + \gamma Q'_{\mu'} \left( s_{i+1}, \mu' \left( s_{i+1}; \theta'_a \right); \theta'_c \right)$ gives the approximate target Q-value that the target network generates upon $N_{batch}$ transitions sampled at random. By differentiating $L(\theta_c)$ w.r.t. $\theta_c$, the gradient is attained:

$$\nabla_{\theta_c} L(\theta_c) \approx \frac{2}{N_{batch}} \sum_{i=1}^{N_{batch}} \left[ \left( Q_\mu(s_i, \mu(s_i; \theta_a); \theta_c) - y_i \right) \times \nabla_{\theta_c} Q_\mu(s_i, \mu(s_i; \theta_a); \theta_c) \right]. \tag{2.19}$$

The training-critic network parameter, $\theta_c$, is updated using the stochastic gradient descent method [24].

The target-actor and target-critic networks are updated based on the training-actor and training-critic networks:

$$\theta'_a \leftarrow \tau_a \theta_a + (1 - \tau_a)\theta'_a,$$
$$\theta'_c \leftarrow \tau_c \theta_c + (1 - \tau_c)\theta'_c, \tag{2.20}$$

where $\tau_a$ and $\tau_c$ are the decaying rates for the training-actor and training-critic networks, respectively.

**Twin Delayed DDPG (TD3)**

TD3 is one of the latest extensions of DDPG and consists of a training network and a target network, where the training network is made of a training-actor and two training-critic networks, and the target network comprises a target-actor and two target-critic networks, as shown in Fig. 2.4. TD3 addresses the Q-value overestima-

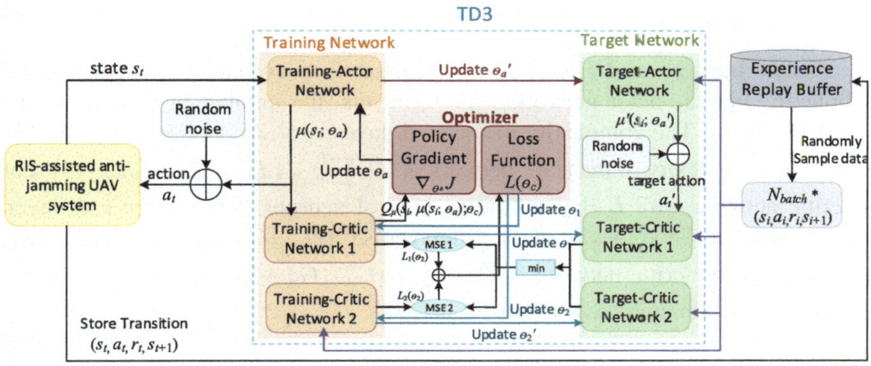


**Fig. 2.4** TD3-based framework with an actor network comprising an actor and a target actor, and a critic network comprising two critics and two target critics. The experience replay buffer gives batches of samples of state transitions for training and updating the networks

tion problem of the DDPG algorithm by incorporating *three improvements* over the classical DDPG model, namely, *clipped double-Q learning*, *target policy smoothing*, and *delayed policy update* [25, 26].

- *Clipped double-Q learning*: TD3 contains two training-critic and target-critic networks to produce two Q-values. The lesser of the two is used to evaluate the target Q-value in the Bellman error loss function. Specifically, $Q_\mu(s_t, a_t; \theta_c)$ in the DDPG is replaced by $Q_\mu(s_t, a_t) = \min\{Q_1(s_t, a_t; \theta_1),$ $Q_2(s_t, a_t; \theta_2)\}$ in the TD3.
- *Target policy smoothing*: TD3 perturbs actions produced by the target-actor network (i.e., "target action") with noises. It smooths the corresponding Q-function values to enhance the resistance of the policy against erroneous Q-functions. The smoothed target action is written as

$$a_t' = \text{clip}\left(\mu'\left(s_{t+1}; \theta_a'\right) + \text{clip}\left(\epsilon', -\sigma_m^2, \sigma_m^2\right), a_{\min}, a_{\max}\right). \tag{2.21}$$

  Here, the noise $\epsilon'$ is taken at random from a Gaussian distribution with zero mean and variance $\sigma_a^2$, i.e., $\epsilon' \sim \mathcal{N}(0, \sigma_a^2)$, and $\sigma_m^2$ is the maximum exploration noise supported by the environment.
- *"Delayed" policy update*: The training-actor and target-actor networks (i.e., policies) are updated less frequently than the training-critic and target-critic networks. For example, it was recommended in [25] that the training-actor and target-actor networks be updated after the training-critic and target-critic networks are updated twice in TD3.

In contrast, the DDPG model only contains a training-critic network and a target-critic network, and produces a Q-value. The DDPG model does not add noises toward target actions. Moreover, the classical DDPG model refreshes its train-actor and target-actor networks and train-critic and target-critic networks at the same pace.

**Complexity and Convergence**

We consider the actor and critic networks separately to evaluate the computational complexity. For the actor network with $L_a$ layers, the number of neurons in the $m$-th layer is $J_m$. The complexity of the $m$-th layer is $\mathcal{O}(J_{m-1}J_m + J_m J_{m+1})$ [27]. Hence, the complexity of the actor network is $\mathcal{O}\Big(\sum_{m=2}^{L_a-1}(J_{m-1}J_m + J_m J_{m+1})\Big)$. For the critic network with $L_c$ layers, the number of neurons is $G_n$ in the $n$-th layer. The computational complexity of the $n$-th layer is $\mathcal{O}(G_{n-1}G_n + G_n G_{n+1})$ [27]. The complexity of the critic network is $\mathcal{O}\Big(\sum_{n=2}^{L_c-1}(G_{n-1}G_n + G_n G_{n+1})\Big)$. As a result, the overall complexity of Algorithm 2.1 is [27]

$$\mathcal{O}\Bigg(\sum_{m=2}^{L_a-1}(J_{m-1}J_m + J_m J_{m+1}) + \sum_{n=2}^{L_c-1}(G_{n-1}G_n + G_n G_{n+1})\Bigg).$$

**Summary**

Our approach here is to harness the power of AI to enhance the communication capabilities of UAV-borne IoT platforms in the presence of jamming threats, i.e., a jammer on the ground tries to jam the communication between a UAV and its ground peer. A DRL architecture can be developed to optimize both flight navigation and RIS-assisted jamming cancellation, ultimately maximizing the data rate of the UAV.

The highlight of the approach lies in the utilization of AI algorithms, e.g., DDPG or TD3. The UAV can act as the agent. Its received data rate is used as the state of AI models. The action of the UAV includes its path and the RIS configuration. These advanced AI models are designed to efficiently handle the continuous state and action spaces associated with UAV and RIS configurations. Through exploration and exploitation, the UAV can learn its flight path and RIS configuration solely based on the received data rate, eliminating the need for computationally expensive estimation of the instantaneous CSI.

The significance of this AI-based approach becomes evident when considering the challenges associated with RIS-assisted jamming rejection in UAV communications. Estimating CSI involving an RIS is known to be a difficult task, especially when the CSI involving the RIS and the uncooperative jammer is considered. The DRL architecture overcomes this limitation by allowing the UAV to adaptively modify the wireless transmission environment using the RIS, mitigating jamming power and enhancing intended signals.

## 2.3.2 *AI for RIS-Based Radio Surveillance*

### 2.3.2.1 System Model

As illustrated in Fig. 2.2, we consider an RIS-assisted, UAV-based surveillance system, where a fixed-wing UAV is employed to monitor and wiretap a suspicious radio link on the ground with the assistance of a benign RIS. The suspicious transmitter (Tx) and the receiver (Rx) are apart for $d$ m. The UAV and the suspicious Tx/Rx all have a single antenna for illustration convenience. The RIS is equipped with a smart controller to receive the instruction from the UAV for its configuration [17].

**Threat Model**

The threat model considered is that the suspicious Tx transmits messages that potentially contain plans or evidence of crimes to its intended Rx in a point-to-point communication mode. This can circumvent typical wiretapping techniques. In this case, radio surveillance is critical to detect crime, determine transmitter identities, or gather evidence for criminal prosecution. The suspicious Tx can be equipped with anti-reconnaissance capability, e.g., cameras. Therefore, it is important for the monitor, i.e., a UAV, to stay unnoticed by keeping its distance from the Tx. For this reason, the RIS-assisted, UAV-based surveillance system has the potential, where an RIS augments the radio propagation environment to allow the UAV to eavesdrop on the suspicious link and keep its distance.

On the other hand, the UAV needs to maintain an active communication link with the RIS to send configurations to the RIS in the described system. However, this may be susceptible to several security attacks, e.g., launched by the Tx or its other accomplices, including (1) eavesdropping, where an attacker eavesdrops on the UAV-RIS link to obtain the UAV's credentials; (2) replay attack, where the attacker replays the UAV's transmissions to mislead the RIS; (3) spoofing, where the attacker pretends to be the UAV by encoding its messages with the UAV's identity and credentials; and 4) message forgery, where the attacker sends forged messages following the spoofing to take control of the RIS.

A countermeasure to these attacks is strong data encoding and encryption at the UAV and RIS to increase the difficulty in deciphering the UAV-RIS link. The UAV and RIS can frequently and regularly update their encryption keys, coordinated by a trusted third party, e.g., a cellular operator. The UAV can timestamp its messages to prevent replay attacks. The UAV can protect its identity and credentials by preventing eavesdropping and replay attacks. This makes the spoofing and message forgery unlikely. Assume that these security measures are in place. The UAV can have a secure and reliable communication link with the RIS.

**Channel Model**

The suspicious Tx and Rx nodes are at (0, 0, 0) and $\boldsymbol{q}_{\text{Rx}} = [d, 0, 0]^T$ on the ground (i.e., the $x$-$y$ plane). The distances of the Tx-UAV link $d^t_{\text{Tx,U}}$, Tx-RIS link $d_{\text{Tx,R}}$, RIS-UAV link $d^t_{\text{R,U}}$, and RIS-Rx link $d_{\text{R,Rx}}$ are

$$d_{\mathrm{Tx,R}} = \|\boldsymbol{q}_{\mathrm{R}}\|,\ d^t_{\mathrm{Tx,U}} = \|\boldsymbol{q}_t\|,\ \forall t, \tag{2.22a}$$

$$d_{\mathrm{R,Rx}} = \|\boldsymbol{q}_{\mathrm{R}} - \boldsymbol{q}_{\mathrm{Rx}}\|,\ d^t_{\mathrm{R,U}} = \|\boldsymbol{q}_t - \boldsymbol{q}_{\mathrm{R}}\|,\ \forall t. \tag{2.22b}$$

**Tx-Rx Link**. Consider a Rayleigh fading channel for the Tx-Rx link. The channel power gain of the Tx-Rx link is

$$h_{\mathrm{Tx,Rx}} = \sqrt{\rho d_{\mathrm{Tx,Rx}}^{-\kappa_1}}\tilde{h}, \tag{2.23}$$

where $\rho$ is the path loss at the reference distance $d_0 = 1$ m with the path loss exponent $\kappa_1 > 2$, and $\tilde{h}$ is the random scattering component modeled by a zero-mean, unit-variance circularly symmetric complex Gaussian (CSCG) random variable.

**Tx/Rx-RIS Link**. The channel power gains of the Tx-RIS and RIS-Rx links, denoted by $\boldsymbol{h}_{\mathrm{Tx,R}}$ and $\boldsymbol{h}_{\mathrm{R,Rx}} \in \mathbb{C}^{N\times 1}$, are

$$\boldsymbol{h}_{\mathrm{Tx,R}} = \underbrace{\sqrt{\rho d_{\mathrm{Tx,R}}^{-\kappa_2}}}_{\text{path loss}} \underbrace{\left(\sqrt{\frac{\beta_1}{1+\beta_1}}\boldsymbol{h}^{los}_{\mathrm{Tx,R}} + \sqrt{\frac{1}{1+\beta_1}}\boldsymbol{h}^{nlos}_{\mathrm{Tx,R}}\right)}_{\text{array response \& small-scale fading}}; \tag{2.24a}$$

$$\boldsymbol{h}_{\mathrm{R,Rx}} = \underbrace{\sqrt{\rho d_{\mathrm{R,Rx}}^{-\kappa_3}}}_{\text{path loss}} \underbrace{\left(\sqrt{\frac{\beta_2}{1+\beta_2}}\boldsymbol{h}^{los}_{\mathrm{R,Rx}} + \sqrt{\frac{1}{1+\beta_2}}\boldsymbol{h}^{nlos}_{\mathrm{R,Rx}}\right)}_{\text{array response \& small-scale fading}}, \tag{2.24b}$$

where $\kappa_2$ and $\kappa_3$ are the path loss exponents for the Tx-RIS and RIS-Rx links, respectively; $\beta_1$ and $\beta_2$ are the Rician factors; and $\boldsymbol{h}^{nlos}_{\mathrm{Tx,R}}$ and $\boldsymbol{h}^{nlos}_{\mathrm{R,Rx}}$ are the non-LoS (NLoS) components with the variables independently drawn from the zero-mean, unit-variance CSCG distribution.

In (2.24a) and (2.24b), $\boldsymbol{h}^{los}_{\mathrm{Tx,R}}$ and $\boldsymbol{h}^{los}_{\mathrm{R,Rx}}$ are the LoS components:

$$\boldsymbol{h}^{los}_{\mathrm{Tx,R}} = \left[1, \ldots, e^{-j\frac{2\pi d_x}{\lambda}(N_x-1)\phi^x_{\mathrm{Tx,R}}}\right]^T \otimes \left[1, \ldots, e^{-j\frac{2\pi d_y}{\lambda}(N_y-1)\phi^y_{\mathrm{Tx,R}}}\right]^T, \tag{2.25a}$$

$$\boldsymbol{h}^{los}_{\mathrm{R,Rx}} = \left[1, \ldots, e^{-j\frac{2\pi d_x}{\lambda}(N_x-1)\phi^x_{\mathrm{R,Rx}}}\right]^T \otimes \left[1, \ldots, e^{-j\frac{2\pi d_y}{\lambda}(N_y-1)\phi^y_{\mathrm{R,Rx}}}\right]^T, \tag{2.25b}$$

where $d_x$ and $d_y$ are the antenna spacings in the directions of the $x$- and $y$-axes, respectively; $\phi^x_{\mathrm{Tx,R}} = x_{\mathrm{R}}/d_{\mathrm{Tx,R}}$ and $\phi^y_{\mathrm{Tx,R}} = y_{\mathrm{R}}/d_{\mathrm{Tx,R}}$ are the spatial frequencies corresponding to the angles-of-arrivals (AoAs) from the Tx to the RIS; and $\phi^x_{\mathrm{R,Rx}} = (d - x_{\mathrm{R}})/d_{\mathrm{R,Rx}}$ and $\phi^y_{\mathrm{R,Rx}} = y_{\mathrm{R}}/d_{\mathrm{R,Rx}}$ are the spatial frequencies corresponding to the angles-of-departures (AoDs) from the RIS to the Rx in the directions of the $x$- and $y$-axes, respectively [20].

**Tx/RIS-UAV Link**. The PLoS channel can be constructed by considering the probabilities of the LoS and NLoS channels, where the LoS probability of the Tx/RIS-UAV link for a given time slot, $p_b^{los,t}$, $b \in \{\mathrm{Tx, R}\}$, is given by [28]

$$p_{b,\mathrm{U}}^{los,t} = \frac{1}{1 + C_{b,\mathrm{U}} \exp\left(-D_{b,\mathrm{U}}[\theta_{b,\mathrm{U}}^{t} - C_{b,\mathrm{U}}]\right)}, \ \forall b, t. \tag{2.26}$$

Here, $C_{b,\mathrm{U}}$ and $D_{b,\mathrm{U}}$ rely on the propagation environment, and

$$\theta_{b,\mathrm{U}}^{t} = \sin^{-1}(\Delta z_{b,\mathrm{U}}^{t} / d_{b,\mathrm{U}}^{t})$$

is the elevation angle in radii, where $\Delta z_{b,\mathrm{U}}^{t}$ is the altitude difference between the Tx or RIS and the UAV.

The channel power gains of the Tx-UAV and RIS-UAV links, denoted by $h_{\mathrm{Tx,U}}^{t}$ and $\boldsymbol{h}_{\mathrm{R,U}}^{t} \in \mathbb{C}^{N\times 1}$, $\forall t$, are

$$h_{\mathrm{Tx,U}}^{t} = p_{\mathrm{Tx,U}}^{los,t} \sqrt{\rho (d_{\mathrm{Tx,U}}^{t})^{-\kappa_4}} + (1 - p_{\mathrm{Tx,U}}^{los,t}) \sqrt{\rho (d_{\mathrm{Tx,U}}^{t})^{-\kappa_4}} h_{\mathrm{Tx,U}}^{nlos,t}, \tag{2.27a}$$

$$\boldsymbol{h}_{\mathrm{R,U}}^{t} = p_{\mathrm{R,U}}^{los,t} \sqrt{\rho (d_{\mathrm{R,U}}^{t})^{-\kappa_5}} \boldsymbol{h}_{\mathrm{R,U}}^{los,t} + (1 - p_{\mathrm{R,U}}^{los,t}) \sqrt{\rho (d_{\mathrm{R,U}}^{t})^{-\kappa_5}} \boldsymbol{h}_{\mathrm{R,U}}^{nlos,t}, \tag{2.27b}$$

where $\kappa_4$ and $\kappa_5$ are the path loss exponents of the Tx-UAV and RIS-UAV links, respectively; and $h_{\mathrm{Tx,U}}^{nlos,t}$ and $\boldsymbol{h}_{\mathrm{R,U}}^{nlos,t}$ are the NLoS components with the variables independently drawn from the zero-mean, unit-variance CSCG distribution.

In (2.27b), $\boldsymbol{h}_{\mathrm{R,U}}^{los,t}$ is the array response, as given by

$$\boldsymbol{h}_{\mathrm{R,U}}^{los,t} = \left[1, \ldots, e^{-j\frac{2\pi d_x}{\lambda}(N_x - 1)\phi_{\mathrm{R,U},x}^{t}}\right]^T \otimes \left[1, \ldots, e^{-j\frac{2\pi d_y}{\lambda}(N_y - 1)\phi_{\mathrm{R,U},y}^{t}}\right]^T, \ \forall t, \tag{2.28}$$

where $\phi_{\mathrm{R,U},x}^{t} = (x_t - x_{\mathrm{R}})/d_{\mathrm{R,U}}^{t}$ and $\phi_{\mathrm{R,U},y}^{t} = (y_t - y_{\mathrm{R}})/d_{\mathrm{R,}}^{t}$ are the spatial frequencies corresponding to the AoDs from the RIS to the UAV along the $x$- and $y$-axes, respectively.

It is worth pointing out that the UAV does not require the CSI of the channels to and/or from the RIS, i.e., $\boldsymbol{h}_{\mathrm{Tx,R}}$, $\boldsymbol{h}_{\mathrm{R,Rx}}$, and $\boldsymbol{h}_{\mathrm{R,U}}$, to produce its trajectory and configure the RIS. Instead, the UAV only observes the transmit rate of the Tx and measures its own received SNR to learn the UAV trajectory and RIS configuration. This consideration is of practical value because the RIS-reflected channels are difficult to estimate without the cooperation of the Tx. Let alone that the UAV has no access to the CSI of the Tx-Rx link [29].

#### 2.3.2.2 Passive Beamforming for Radio Surveillance

Under the described DRL framework, the UAV only assesses the resolvability of its received signals, and adjusts its flight and the RIS in an attempt to keep the signals resolvable at the UAV. (The received signals are always resolvable at the intended receiver, since the Tx can adjust its transmit rate, adapting to the channel capacity between itself and the Rx.) In other words, the Tx-UAV link needs to sustain a higher channel capacity than the Tx-Rx link, so that the UAV can capture distinguishable and resolvable signal constellations useful for signal detection or cyberanalysis.

A channel outage occurs when the UAV cannot resolve the signal constellations intended for the Rx (because the channel capacity of the Tx-UAV channel is lower than the transmit rate of the Tx-Rx link). To accomplish radio surveillance and prevent a channel outage, the UAV maneuvers its flight and remotely configures the RIS to enhance its own channel from the Tx, or degrade the channel between the Tx and Rx, or both. By adjusting the RIS configuration in real time, the UAV can maintain a high channel capacity for the Tx-UAV link and prevent channel outage, while also ensuring that the Rx can still receive the desired signal.

We design that when the Tx-UAV link has a lower gain than the Tx-Rx link, the RIS adjusts its phase shifts to weaken the Tx-Rx channel and lower the transmit rate of the Tx-Rx link until the UAV can recognize the signal constellations of the Tx node with reasonable resolution. When the Tx-UAV link has a higher gain than the Tx-Rx link, the RIS enhances the Tx-Rx channel as long as the received signals remain resolvable at the UAV. This allows the Tx to send more data to the Rx, and the UAV can shed more insights into the interaction between the Tx and Rx.

The eavesdropping data rate of the UAV at time slot $t$ can be evaluated as

$$R_{\mathrm{U}}^t = \begin{cases} R_{\mathrm{Rx}}^t = \log_2(1+\gamma_{\mathrm{Rx}}^t), \text{ if } \gamma_{\mathrm{U}}^t \geq \gamma_{\mathrm{Rx}}^t; \\ 0, \qquad\qquad\qquad\qquad \text{otherwise.} \end{cases} \tag{2.29}$$

Here, the SNRs at the UAV, $\gamma_{\mathrm{U}}^t$, and the Rx, $\gamma_{\mathrm{Rx}}^t$, are

$$\gamma_{\mathrm{U}}^t = \frac{P_t}{\sigma_{\mathrm{U}}^2}|h_{\mathrm{Tx,U}}^t + (\boldsymbol{h}_{\mathrm{R,U}}^t)^H \boldsymbol{\Theta}_t \boldsymbol{h}_{\mathrm{Tx,R}}|^2, \ \forall t; \tag{2.30a}$$

$$\gamma_{\mathrm{Rx}}^t = \frac{P_t}{\sigma_{\mathrm{Rx}}^2}|h_{\mathrm{Tx,Rx}} + \boldsymbol{h}_{\mathrm{R,Rx}}^H \boldsymbol{\Theta}_t \boldsymbol{h}_{\mathrm{Tx,R}}|^2, \ \forall t, \tag{2.30b}$$

where $P_t$ is the transmit power of the Tx, and $\sigma_{\mathrm{U}}^2$ and $\sigma_{\mathrm{Rx}}^2$ are the variances of the additive white Gaussian noise (AWGN) at the UAV and the Rx, respectively.

The UAV can practically evaluate its SNR, i.e., $\gamma_{\mathrm{U}}^t$ and the transmit rate over the suspicious Tx-Rx link, i.e., $R_{\mathrm{Rx}}^t$.[1] Specifically, the UAV can estimate the CSI-at-the-Rx (CSIR) based on the pilot of the Tx in the same way as the Rx, especially when the Tx, Rx, and UAV are equipped with multiple antennas [30]. With the CSIR, the UAV can estimate its SNR, $\gamma_{\mathrm{U}}^t$, and recover the signal constellations to estimate the transmit rate over the suspicious Tx-UAV link, $R_{\mathrm{Rx}}^t$, if the signal constellations are resolvable at the UAV, i.e., $\gamma_{\mathrm{U}}^t \geq \gamma_{\mathrm{Rx}}^t$. Otherwise, $R_{\mathrm{U}}^t = 0$, as shown in (2.29); the UAV needs to adjust the RIS and maneuver its flight until the constellations become resolvable at the UAV.

[1] The transmit rate $R_{\mathrm{Rx}}^t$ depends on the SNR at the Rx, i.e., $\gamma_{\mathrm{Rx}}^t$; see (2.29). Yet, the UAV does not need to explicitly know $\gamma_{\mathrm{Rx}}^t$ in the described system.

The new model can be extended under imperfect CSIR (i.e., at the intended Rx and the UAV). Define the effective channels from the suspicious Tx to the UAV and the Rx at time slot $t$ as $h^t_{\text{eff,U}} = h^t_{\text{Tx,U}} + (\boldsymbol{h}^t_{\text{R,U}})^H \boldsymbol{\Theta}_t \boldsymbol{h}^t_{\text{Tx,R}}$ and $h^t_{\text{eff,Rx}} = h^t_{\text{Tx,Rx}} + (\boldsymbol{h}^t_{\text{R,Rx}})^H \boldsymbol{\Theta}_t \boldsymbol{h}^t_{\text{Tx,R}}$, respectively. A Minimum Mean-Square Error (MMSE) estimator is often adopted at the Rx [30]. The distribution of channel estimation errors is typically agnostic to the channel distribution. The estimation errors, denoted by $\Delta h^t_{\text{eff,U}}$ and $\Delta h^t_{\text{eff,Rx}}$, are Gaussian, i.e., $\Delta h^t_{\text{eff,U}}$, $\Delta h^t_{\text{eff,Rx}} \in \mathcal{CN}(0, \epsilon_{\text{eff}})$, where $\mathbb{E}[|\Delta h^t_{\text{eff,Rx}}|^2] = \mathbb{E}[|\Delta h^t_{\text{eff,U}}|^2] = \epsilon_{\text{eff}}$ only depends on the noises at the Rx/UAV and is independent of the RIS configuration $\boldsymbol{\Theta}_t$ and the channels. The SNRs/SINRs at the Rx and the UAV in (2.29) can be rewritten as

$$\gamma^t_{\text{U}} = \frac{P_t |\hat{h}^t_{\text{eff,U}}|^2}{P_t \mathbb{E}[|\Delta h^t_{\text{eff,U}}|^2] + \sigma^2_{\text{U}}}, \ \forall t; \tag{2.31a}$$

$$\gamma^t_{\text{Rx}} = \frac{P_t |\hat{h}^t_{\text{eff,Rx}}|^2}{P_t \mathbb{E}[|\Delta h^t_{\text{eff,Rx}}|^2] + \sigma^2_{\text{Rx}}}, \ \forall t, \tag{2.31b}$$

where $\hat{h}^t_{\text{eff,U}} = h^t_{\text{eff,U}} + \Delta h^t_{\text{eff,U}}$ and $\hat{h}^t_{\text{eff,Rx}} = h^t_{\text{eff,Rx}} + \Delta h^t_{\text{eff,Rx}}$ are the estimated effective channels.

At time slot $t$, the eavesdropping data rate of the UAV is

$$R^t_{\text{U}} = \begin{cases} R^t_{\text{Rx}} = \log_2 \left( \dfrac{P_t |\hat{h}^t_{\text{eff,Rx}}|^2}{P_t \epsilon_{\text{eff}} + \sigma^2_{\text{Rx}}} \right), & \text{if } \gamma^t_{\text{U}} \geq \gamma^t_{\text{Rx}}; \\ 0, & \text{otherwise,} \end{cases} \tag{2.32}$$

which shows that the channel estimation errors (at the Rx) would result in reduced SNR/SINR and an increased requirement of the transmit power of the Tx. The UAV could underestimate the transmit rate due to less accurately recovered signal constellations. The UAV would configure the RIS to further degrade the Tx-Rx link and/or enhance the Tx-UAV link until the signal constellations become resolvable at the UAV, as in the case of perfect CSIR.

#### 2.3.2.3 TD3 Framework for UAV Navigation and RIS Configuration

This section presents a new TD3-based method under the actor-critic framework to solve problem (2.10). Our method learns to configure the RIS and navigate the UAV (including heading and acceleration) based on its observed transmit rate of the suspicious link, suppressing the need for precise CSI, including the Tx-Rx link and the RIS reflecting channels.

**State, Action, and Reward**

The UAV trajectory (i.e., waypoints) is a Markov decision process (MDP). The RIS configuration depends on the instantaneous position of the UAV. We can interpret problem (2.10) as an MDP with its state, action, and reward defined below.

***State Space*** $\mathcal{S}$: At time slot $t$, the system state $s_t \in \mathcal{S}$ is made of the relative position of the UAV with regard to its final location, $\boldsymbol{q}_t - \boldsymbol{q}_F$; the UAV's velocity $\boldsymbol{V}_t$; the SNRs at the UAV $\gamma_{\mathrm{U}}^t$ and the Rx node $\gamma_{\mathrm{Rx}}^t$,[2]; and the distance between the Tx-UAV link $d_{\mathrm{Tx,U}}^t$ at the $t$-th time step, i.e., $s_t = \{\boldsymbol{q}_t - \boldsymbol{q}_F, \boldsymbol{V}_t, \gamma_{\mathrm{U}}^t, \gamma_{\mathrm{Rx}}^t, d_{\mathrm{Tx,U}}^t\}$.

***Action Space*** $\mathcal{A}$: The action space collects all possible actions, i.e., $a_t \in \mathcal{A}$. At the $t$-th time step, the action, $a_t$, consists of two parts: The reflecting coefficients of the RIS, i.e., $\{\theta_n^t\}_{n\in\mathcal{N}}$, and the acceleration of the UAV, i.e., $\boldsymbol{A}_t := [A_{xt}, A_{yt}, A_{zt}]^T$. The action $a_t$ has $(N+3)$ dimensions. The UAV's acceleration is constrained by (2.1d)–(2.1f). Its waypoints $\boldsymbol{q}_t$ and velocities $\boldsymbol{V}_t$ depend on the accelerations, i.e., (2.1a)–(2.1c).

We exploit the normalized representation for both the reflecting coefficients of the RIS and the acceleration of the UAV, i.e., $\boldsymbol{A}_t = A_{\max}\boldsymbol{\lambda}_A$ and $\theta_n^t = \pi\lambda_\theta(n)$, where $\boldsymbol{\lambda}_A = [\lambda_A(0), \lambda_A(1), \lambda_A(2)]$ and $n \in \mathcal{N}$. Here,

$$-1 \leq \lambda_A(0), \lambda_A(1), \lambda_A(2), \lambda_\theta(n) \leq 1.$$

***Reward*** $r_t$: The reward function is designed based on the objective and constraints in (2.10) to evaluate the action $a_t$ at each time slot. The reward function has three parts:

1. *Eavesdropping Rate Maximization*. The reward for maximizing the amount of eavesdropped data while disguising the UAV's intention of eavesdropping by keeping a predefined distance, i.e., $D_{\mathrm{cov}}$, from the Tx node at any time slot $t$, is defined as

$$r_t^{\{er\}} = \begin{cases} \zeta_{er} R_{\mathrm{U}}^t, & \text{if } \hat{d}_{\mathrm{Tx,U}}^t \geq D_{\mathrm{cov}}; \\ 0, & \text{otherwise,} \end{cases} \tag{2.33}$$

   where $\zeta_{er}$ is a positive constant used to tune the relative importance of this part in the reward function.

2. *Destination Approaching*. The reward for destination approaching is designed as

$$r_t^{\{dr\}} = \zeta_{dr}\left(d_F^{t-1} - d_F^t\right), \tag{2.34}$$

   where $\left(d_F^{t-1} - d_F^t\right)$ is the reduced distance from the UAV to the final location after time slot $t$; $d_F^{t-1} = \|\boldsymbol{q}_{t-1} - \boldsymbol{q}_F\|$ and $d_F^t = \|\boldsymbol{q}_t - \boldsymbol{q}_F\|$ are the distances from the UAV to the final location at slots $(t-1)$ and $t$, respectively; and $\zeta_{dr}$ is a positive constant to adjust the relative importance of this part in the reward function.

   With the reward $\zeta_{dr}\left(d_F^{t-1} - d_F^t\right)$, the UAV is incentivized to get closer to the final location at each time step. Moreover, $\zeta_{dr}\left(d_F^{t-1} - d_F^t\right)$ provides a positive

[2] The UAV estimates the rate of the Tx, $\log_2(1+\gamma_{\mathrm{Rx}}^t)$, and recover $\gamma_{\mathrm{Rx}}^t$.

reward if the distance from the UAV to the final location is shorter at the $t$-th time step than it was at slot $(t-1)$. Otherwise, a penalty occurs.

When $t = T_w$, we set a reward $r_t^{\{arr\}}$ to indicate whether the UAV reaches the final location:

$$r_t^{\{arr\}} = \begin{cases} 0, \ t = 1, \ldots, T_w - 1; \\ \zeta_{arr}(t) r_{arr} + (1 - \zeta_{arr}(t)) \, r_{narr}, \ t = T_w, \end{cases} \tag{2.35}$$

where $r_{arr}$ is a positive constant to motivate the arrival of the UAV within the desired region centered at the final location, and $r_{narr}$ is a negative constant to penalize if the UAV does not arrive at the destination. $\zeta_{arr}(t) = 1$ if the UAV reaches the desired final location at the final time step (i.e., $t = T_w$). Otherwise, $\zeta_{arr}(t) = 0$.

3. *Constraints.* We also set penalties to punish infeasible actions that cannot satisfy constraints (2.1e) and (2.1f):

$$\begin{aligned} r_t^{\{v\}} &= \zeta_v(t) \cdot r_v, \ \text{if } \|\boldsymbol{V}_t\| \in [V_{\min}, V_{\max}]; \\ r_t^{\{z\}} &= \zeta_z(t) \cdot r_z, \ \text{if } z_t \in [z_{\min}, z_{\max}], \end{aligned} \tag{2.36}$$

where $\zeta_v(t)$ is the velocity constraint indicator with $\zeta_v(t) = 1$ indicating that the UAV's velocity violates constraint (2.1e) and $\zeta_v(t) = 0$ indicating otherwise; and $\zeta_z(t)$ is the altitude constraint indicator with $\zeta_z(t) = 1$ indicating that the UAV's altitude violates constraint (2.1f) and $\zeta_z(t) = 0$ indicating otherwise.

The overall reward function is

$$r_t = r_t^{\{er\}} + r_t^{\{dr\}} + r_t^{\{arr\}} + r_t^{\{v\}} + r_t^{\{z\}} \tag{2.37}$$

***Policy***: A mapping from the state space $\mathcal{S}$ to the action space $\mathcal{A}$ is known as a policy, $\mu : \mathcal{S} \to \mathcal{A}$, which is a distribution $\mu(a|s) = \Pr(a_t = a|s_t = s)$ over $s \in \mathcal{S}$. A policy $a_t = \mu(s_t)$ maps state $s_t$ to a feasible action. The agent selects the policy maximizing the accumulated reward $R_t = \sum_{n=t}^{N} \gamma^{n-t} r_t$ with $\gamma \in [0, 1]$ being a discount factor. Given state $s_t$, action $a_t$, and policy $\mu(\cdot \cdot)$, $R_t$ is evaluated by an action-value function, i.e., the Q-function, as given by

$$Q_\mu(s_t, a_t) = \mathbb{E}_\mu[R_t|s_t, a_t], \tag{2.38}$$

which satisfies the Bellman Expectation Equation:

$$Q_\mu(s_t, a_t) = \mathbb{E}_{r_t, s_{t+1} \sim \mathcal{E}}\left[r_t + \gamma \mathbb{E}_{a_{t+1} \sim \mu}\left[Q_\mu(s_{t+1}, a_{t+1})\right]\right], \tag{2.39}$$

where $\mathcal{E}$ denotes the environment that the agent interacts with.

***Experience***: An experience is defined as $e_t = (s_t, a_t, r_t, s_{t+1})$ and stored in an experience replay buffer $\boldsymbol{R}$.

**Algorithm 2.1:** Proposed TD3 model for solving (2.10).

**Initialization**: Randomly initialize the actor $\mu$ and the two critics $Q_1$ and $Q_2$ with parameters $\theta_a$, $\theta_1$, and $\theta_2$, the target-actor $\mu'$ and two target-critics $Q_1'$ and $Q_2'$ with parameters $\theta_a' \leftarrow \theta_a$, $\theta_1' \leftarrow \theta_1$, and $\theta_2' \leftarrow \theta_2$, the experience replay buffer $\boldsymbol{R}$, and the channel allocation and modulation-coding selection $\boldsymbol{\eta}_0$.

Initialize the relative locations and velocities from the UAV to the final location, and estimate the eavesdropping rate of the UAV as the initial state $s_0$.

**for** $t = 1, \ldots, T_w$ **do**

  Pick action $a_t = \text{clip}\Big(\mu\left(s_t; \theta_a\right) + \epsilon, a_{\min}, a_{\max}\Big)$.

  Execute action $a_t$, receive the reward $r_t$, perceive a new state $s_{t+1}$, and store transition $(s_t, a_t, r_t, s_{t+1})$ in $\boldsymbol{R}$.

  Randomly sample $N_{batch}$ historical transitions $(s_i, a_i, r_i, s_{i+1})$ from $\boldsymbol{R}$.

  Update the target action after target policy smoothing based on the sampled transitions: $a_t' = \text{clip}\Big(\mu'\left(s_{t+1}; \theta_a'\right) + \text{clip}\left(\epsilon', -\sigma_m^2, \sigma_m^2\right), a_{\min}, a_{\max}\Big)$.

  Update the target Q-value: $y_i = r_i + \gamma \min\Big(Q_1'(s_{i+1}, a_i'; \theta_1'), Q_2'(s_{i+1}, a_i'; \theta_2')\Big)$.

  Calculate the loss function based on (2.44), and update the two critics by (2.45).

  **if** mod $(t, 2) = 0$ **then**

    Update the actor based on (2.42), and the target-actor and the two target-critics by (2.46).

In terms of convergence, the TD3-based algorithm satisfies the following conditions: (i) the network parameters $\theta$ and $\theta'$ (with the subscripts omitted for brevity) are upper bounded since they are sequentially compact following the Arzela-Ascoli theorem [31]; (ii) the state and action spaces are compact, as the sampled states and actions are bounded by the UAV's maximum speed and acceleration; (iii) the reward function (16) is continuous; and (iv) the training networks are feedforward FCNNs with twice continuously differentiable activation functions. According to [32, Lemma 2], the algorithm can asymptotically converge as the number of iterations tends to infinity since we adopt a sequence of square summable learning rates, i.e., $\sum_t \eta_a(t) = \infty$ and $\sum_t \eta_a^2(t) < \infty$. Here, $t$ indicates a time step. $\eta_a(t)$ is the learning rate of the actor network at the $t$-th time step.

**TD3-based Algorithm**

Algorithm 2.1 summarizes the proposed TD3-based algorithm. Based on the UAV's current state, an action $a_t$ is produced by the actor following the DPG theorem [21]. We use the tanh$(\cdot)$ activation function to output $\lambda_A$ and $\lambda_\theta$. A random exploration noise is added to the action to balance the exploration of new actions and the exploitation of known actions. The output action is

$$a_t = \text{clip}\left(\mu(s_t; \theta_a) + \epsilon, a_{\min}, a_{\max}\right), \tag{2.40}$$

where the noise $\epsilon$ is randomly sampled from a zero-mean Gaussian distribution with variance $\sigma_e^2$, i.e., $\epsilon \sim \mathcal{N}(0, \sigma_e^2)$; clip$(\cdot)$ is a clipping function to limit the actions within the upper and lower bounds of the actions, denoted by $a_{\max}$ and $a_{\min}$, respectively. The UAV executes $a_t$, refreshes its state, collects its reward $r_t$, and transits to the next state $s_{t+1}$. The transition $(s_t, a_t, r_t, s_{t+1})$ is recorded in the experience replay buffer $\boldsymbol{R}$.

Due to the overfitting issue of the deterministic policies [21], the target Q-value produced by the deterministic policy is susceptible to the inaccuracies induced by the Q-function estimation errors. *Target policy smoothing*, a regularization strategy for Q-function value learning [33] is used to reduce the inaccuracies. With $N_{batch}$ transitions randomly sampled from $\boldsymbol{R}$, the target action after the target policy smoothing is

$$a_t' = \text{clip}\Big(\mu'\left(s_{t+1}; \theta_a'\right) + \text{clip}\left(\epsilon', -\sigma_m^2, \sigma_m^2\right), a_{\min}, a_{\max}\Big), \tag{2.41}$$

where the noise $\epsilon'$ is randomly sampled from a zero-mean Gaussian distribution with variance $\sigma_a^2$, i.e., $\epsilon' \sim \mathcal{N}(0, \sigma_a^2)$. $\epsilon'$ is clipped to be within $[-\sigma_m^2, \sigma_m^2]$ with $\sigma_m^2$ pre-specified.

Given the probability distribution of the parameter $\theta_a$, i.e., $J(\theta_a)$, the actor network is updated in the direction that the estimation improves the strategy fastest. With the $N_{batch}$ transitions from $\boldsymbol{R}$, $\nabla_{\theta_a} J(\theta_a)$ is approximated by

$$\nabla_{\theta_a} J(\theta_a) \approx \frac{1}{N_{batch}} \sum_{i=1}^{N_{batch}} \left[\nabla_{\theta_a} \mu(s_i) \nabla_a Q_\mu(s_i, a; \theta_c)|_{a=\mu(s_i)}\right]. \tag{2.42}$$

The model parameter of the actor, i.e., $\theta_a$, is updated by using the gradient descent method [24]

$$\begin{aligned} \theta_a &\leftarrow \theta_a + \eta_a \nabla_{\theta_a} J(\theta_a) \\ &\approx \theta_a + \frac{\eta_a}{N_{batch}} \sum_{i=1}^{N_{batch}} \left[\nabla_{\theta_a} \mu(s_i) \nabla_a Q_\mu(s_i, a; \theta_c)|_{a=\mu(s_i)}\right], \end{aligned} \tag{2.43}$$

where $\eta_a$ is the learning rate of the actor network $\mu(s_t; \theta_a)$ [22]; $\nabla_{\theta_a} \mu(s)$ is the gradient of the actor $\mu(s)$ with respect to the parameter $\theta_a$; and $\nabla_a Q_\mu(s_t, a; \theta_a)$ is the gradient of $Q_\mu(s_t, a; \theta_a)$ with respect to the action $a$.

The mean square error (MSE)-based loss function, $L_k(\theta_k)$, of the two critics is approximately evaluated as

$$L_k(\theta_k) \approx \frac{1}{N_{batch}} \sum_{i=1}^{N_{batch}} \left[\left(Q_k\left(s_i, a_i\right) - y_i\right)^2\right],\ k = 1, 2, \tag{2.44}$$

where $y_i = r_i + \gamma \min\Big(Q_1'(s_{i+1}, a_i'; \theta_1'), Q_2'(s_{i+1}, a_i'; \theta_2')\Big)$ is the approximate target Q-value produced by the target network based on $N_{batch}$ randomly sampled transitions; $\gamma$ is a discount factor; and $\theta_1'$ and $\theta_2'$ are decayed copies of $\theta_1$ and $\theta_2$, respectively.

The gradient of $L_k(\theta_k)$ is given by

$$\nabla_{\theta_k} L_k(\theta_k) \approx \frac{1}{N_{batch}} \sum_{i=1}^{N_{batch}} \Big[ (Q_\mu (s_i, \mu(s_i; \theta_a); \theta_k) \\ - y_i) \nabla_{\theta_k} Q_\mu (s_i, \mu(s_i; \theta_a); \theta_k) \Big], \ k = 1, 2. \tag{2.45}$$

The parameters of the two critics, i.e., $\theta_1$ and $\theta_2$, are updated by utilizing the stochastic gradient descent method [24].

According to the *"delayed" policy updates*, the target actor and the two target critics evolving from the actor and critics are updated every two iterations by running the Polyak Averaging [25], as given by

$$\theta'_a \leftarrow \rho_\tau \theta_a + (1-\rho_\tau)\theta'_a; \theta'_k \leftarrow \rho_\tau \theta_k + (1-\rho_\tau)\theta'_k, \forall k, \tag{2.46}$$

where $\rho_\tau$ is the decaying rate of both actor and critic networks.

The model is trained iteratively until the convergence of the cumulative reward.

**Complexity and Convergence**

To evaluate the computational complexity, we consider the actor and critic networks separately. For the actor network with $L_a$ layers, the number of neurons in the $m$-th layer is $J_m$. The complexity of the $m$-th layer is $\mathcal{O}(J_{m-1}J_m + J_m J_{m+1})$ [27]. Hence, the complexity of the actor network is $\mathcal{O}\left(\sum_{m=2}^{L_a-1}(J_{m-1}J_m + J_m J_{m+1})\right)$. For the critic network with $L_c$ layers, the number of neurons is $G_n$ in the $n$-th layer. The computational complexity of the $n$-th layer is $\mathcal{O}(G_{n-1}G_n + G_n G_{n+1})$ [27]. The complexity of the critic network is $\mathcal{O}\left(\sum_{n=2}^{L_c-1}(G_{n-1}G_n + G_n G_{n+1})\right)$. As a result, the overall complexity of Algorithm 2.1 is [27]

$$\mathcal{O}\left(\sum_{m=2}^{L_a-1}(J_{m-1}J_m + J_m J_{m+1}) + \sum_{n=2}^{L_c-1}(G_{n-1}G_n + G_n G_{n+1})\right).$$

In terms of convergence, the TD3-based algorithm satisfies the following conditions: (i) the network parameters $\theta$ and $\theta'$ (with the subscripts omitted for brevity) are upper bounded since they are sequentially compact following the Arzela-Ascoli theorem [31]; (ii) the state and action spaces are compact, as the sampled states and actions are bounded by the UAV's maximum speed and acceleration; (iii) the reward function (16) is continuous; and (iv) the training networks are feedforward FCNNs with twice continuously differentiable activation functions. According to [32, Lemma 2], the algorithm can asymptotically converge as the number of iterations tends to infinity since we adopt a sequence of square summable learning rates, i.e., $\sum_t \eta_a(t) = \infty$ and $\sum_t \eta_a^2(t) < \infty$. Here, $t$ indicates a time step. $\eta_a(t)$ is the learning rate of the actor network at the $t$-th time step.

**Summary**

While UAVs have been demonstrated to be an effective means for radio surveillance, they are often still restrained by accessibility to the environments where suspicious transmitters are. The approach discussed in this section combines UAVs with legitimate RISs to enhance the radio propagation channel in favor of the UAVs. The approach involves sequential decision-making with a non-convex and non-smooth objective, which is challenging for traditional optimization methods [26]. DRL, e.g., DDPG and TD3, can be applied to navigate the UAVs and configure the RISs, thereby capturing high-resolution radio fingerprints of suspicious transmitters. A UAV can learn its action, i.e., its flight path and the RIS configuration, based on the transmit rates of the suspicious transmitter that can be readily observed and measured at the UAVs (if the channel is good between the suspicious transmitter and UAV).

The development of a DRL framework allows the UAV to learn its actions solely based on observed transmit rates of the suspicious transmitter. The DRL model considers the mobility constraint of the UAV and maximizes the eavesdropping rate by defining appropriate actions and rewards. This approach enables efficient surveillance and eliminates the need for complex CSI estimation involving the RIS and transmitter-receiver pairs.

## 2.4 Performance Evaluation

This section offers Python numerical results to verify the advantages of applying DRL to the new RIS-empowered UAV security. We consider Rayleigh fading channels for the communication link between the transmitter (Tx) and receiver (Rx), and Rician fading channels for the communication links among the Tx, Rx, UAV, Jammer, and RIS. The locations of each network element will be specified later for each individual experiment.

Both DDPG- and TD3-based algorithms are implemented, and the architecture of the algorithms is delineated in Fig. 2.3. The configurations of the algorithms are provided below:

- *DDPG-based algorithm*: The actor networks are implemented by fully connected neural networks (FCNNs) with three hidden layers and their learning rates are $10^{-4}$. The first layer has 64 neurons. The second layer has 128 neurons. The third layer has 64 neurons. The output layer of the actor networks uses the tanh(·) activation function to constrain the output actions not to exceed $[-\pi, \pi)$ for the RIS configuration and $[-2, 2]$ m/s$^2$ for the UAV control. The critic networks use the FCNNs with two hidden layers and their learning rates are $10^{-3}$. Both hidden layers utilize the Rectified Linear Unit (ReLU) activation functions. The first layer has 64 neurons. The second layer has 128 neurons. The additive noise $\mathcal{N}$ to train the DDPG actor policy is sampled from the complex Gaussian noise that has zero mean and variance 0.2.
- *TD3-based algorithm*: This algorithm is extended from the above DDPG network by duplicating the training-critic and target-critic networks of the DDPG network.

The exploration noise added when training the TD3 actor is sampled from the complex Gaussian noise that has zero mean and variance 0.2 second, as in the DDPG network. The policy noise added to flatten the target actor is sampled from the complex Gaussian noise having zero mean and variance of 0.2. The exploration noise does not exceed 0.5. The update frequency of the actor networks is 2.

The DDPG- and TD3-based networks are built on Python 3.9 PyTorch v.1.8.1, and trained on a server that has an Nvidia Tesla P100 SXM2 16GB GPU.

As discussed in Section I, no existing works have addressed the RIS-assisted jamming rejection or radio surveillance for a UAV-borne IoT platform. For this reason, it may not be adequate to compare the described approach with those existing works. With due diligence, we come up with two baselines to further verify the merits of the approach:

***Baseline 1***: In this scheme, we assume that the instantaneous CSI of all involved links is perfectly known for the configuration of the RIS, referred to as "*RDU*." Accordingly, the signals from various paths are added coherently at the UAV to maximize the received signal power and, in turn, the achievable communication rate. The phase shifts of the RIS are configured to align the phases of the Tx-UAV link and the RIS-UAV link, at the UAV.

***Baseline 2***: It is also prudent to compare with the TD3-based scheme without the assistance of the RIS (i.e., $N = 0$), referred to as "*TD3 without RIS*," where the action space only consists of the UAV accelerations. This baseline helps to reveal the gain stemming from using the RIS.

### *2.4.1 Jamming Rejection*

We evaluate the new DDPG-based and TD3-based algorithms for $N = 20$ and 40 at the RIS, as well as the baseline schemes for $N = 20$ at the RIS for comparison. This is consistent with the existing works on RIS-assisted, multi-user communications, e.g., [6], which have typically considered fewer than 20 RIS elements due to the high complexity of RIS optimization. The locations of the Tx and jammer are $[0, 0, 0]^T$ m and $[-50, 0, 0]^T$ m. The initial and final locations of the UAV are $\boldsymbol{q}_0 = [0, -100, 5]^T$ m and $\boldsymbol{q}_F = [150, 100, 100]^T$ m. The RIS has $N = 5 \times 4 = 20$ or $N = 5 \times 8 = 40$ reflecting elements, and the reference point of the RIS is $[50, 50, 30]^T$ m. The scheduling horizon is $T = 30$ seconds with $\delta = 0.1$ second per time slot. 2,000 testing episodes are conducted, each consisting of 300 slots. When testing, no exploration noise is imposed. The described algorithms and baselines are evaluated in terms of the 3D UAV trajectory and received data rate.

Figure 2.5 shows the 3D trajectory of the UAV. These figures demonstrate that the UAV is able to adapt its control policy to the anti-jamming communication and successfully reach the destination. In general, the TD3 converges faster and more smoothly than the DDPG. Yet, it undergoes less smooth changes in the per-episode reward when the action space is smaller, i.e., $N = 20$. This is because the TD3 has

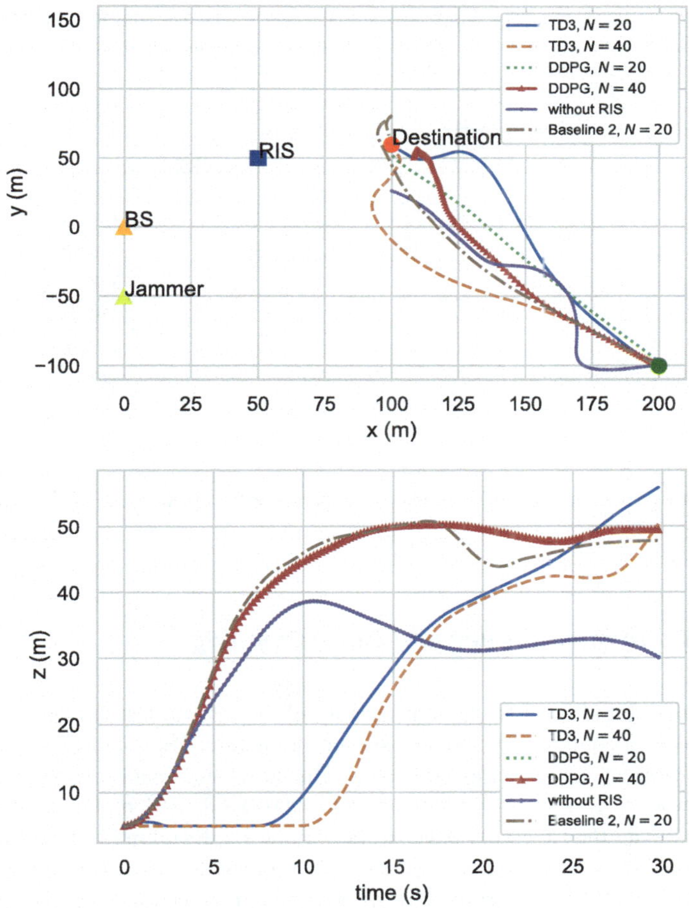


**Fig. 2.5** The 3D UAV trajectory for anti-jamming, where the green and red dots are the UAV initial location and its expected destination, the orange and yellow triangles denote the locations of the Tx and jammer, and the blue square is the RIS reference point

two critic networks for both the training and target networks, which could incur higher complexity and more randomness when training, especially when the action space is smaller. In contrast, DDPG is suitable for training tasks that are less complex and have relatively smaller action spaces.

Table 2.1 provides the data rate of the UAV under the described and baseline schemes. It is seen that the data rate increases with the number of RIS elements. Compared with the scheme without RIS, the described TD3 and DDPG schemes with RIS show an increase in data rate of 7% and 2%, respectively, when $N = 20$. Additionally, the described TD3 scheme achieves a higher data rate than the described DDPG scheme, particularly for larger numbers of RIS elements. This is attributed

**Table 2.1** Comparison of anti-jamming capability between the DDPG- and TD3-based algorithms with the baseline schemes, where RDU is based on an ideal assumption with the instantaneous perfect CSI of all channels available to the UAV

| Scheme | Number of RIS elements (N) | Average rate (bit/Hz/s) | Total rate (bit/Hz/s) |
|---|---|---|---|
| TD3 | N = 20 | 2.46 | 366.51 |
| | N = 40 | 2.48 | 369.07 |
| DDPG | N = 20 | 2.34 | 348.85 |
| | N = 40 | 2.41 | 358.61 |
| RDU | N = 20 | 2.35 | 350.05 |
| | N = 40 | 2.53 | 377.38 |
| TD3, without RIS | N = 0 | 2.30 | 343.27 |

to the faster convergence and better convergent control policy of TD3 compared to DDPG.

### 2.4.2 *Radio Surveillance and Eavesdropping*

We evaluate the radio surveillance capability of the schemes in terms of the eavesdropping success probability and the eavesdropping rate. The channel fading is independently generated among the 2,000 tests. The initial and final locations of the UAV monitor are $\boldsymbol{q}_0 = [200, -100, 5]^T$ m and $\boldsymbol{q}_F = [100, 60, 50]^T$ m. For each time slot, we also evaluate the (average) eavesdropping rate by averaging the eavesdropping rates of the 2,000 tests at the time slot. Eavesdropping is successful at time slot $t$ if the SNR received by the UAV, denoted by $\gamma_{\mathrm{U}}^t$, is greater than the received SNR at the Rx, denoted by $\gamma_{\mathrm{Rx}}^t$; i.e., $\gamma_{\mathrm{U}}^t \geq \gamma_{\mathrm{Rx}}^t$; in other words, the UAV can resolve the constellation of its eavesdropped signals. At any time slot $t$, the per-slot eavesdropping success probability is the probability that the UAV can successfully eavesdrop on the Tx at the time slot, i.e., $\Pr\{\gamma_{\mathrm{U}}^t \geq \gamma_{\mathrm{Rx}}^t\}$ as per $t$.

We plot the tested 3D UAV trajectories of the control policy learned by the described and the baseline schemes in Fig. 2.6a, where the TD3 algorithms under $N = 20$ and $N = 40$ are shown by the blue solid line and the orange dash line, respectively, the RDU scheme under $N = 20$ is shown by the green dash-dot line, and the DDPG algorithm under $N = 20$ is shown by the red dot line. We also plot the projection of the 3D UAV trajectory in the $x$-$y$ plane, as depicted in Fig. 2.6b. As revealed in the figures, the UAV can quickly adapt its control policy (i.e., the acceleration) to the surveillance mission and fly to the final location. In particular, under the TD3 algorithm when $N = 40$, the UAV first flies close to the RIS, and then to the final destination.

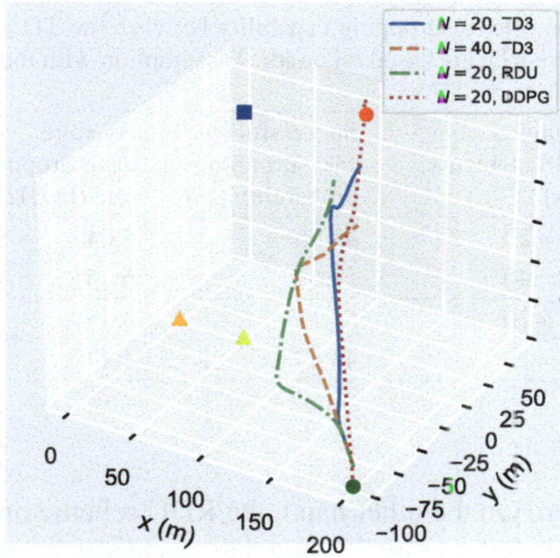


(a) The 3D UAV trajectory for radio surveillance.

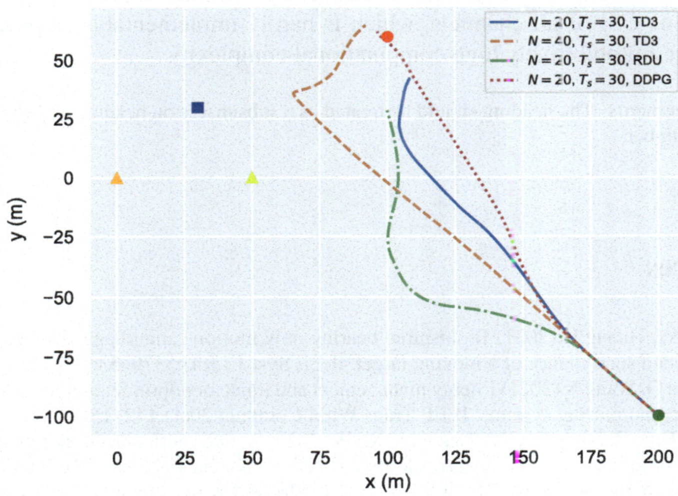


(b) Projection of 3D UAV trajectory on the $x, y$-plane.

**Fig. 2.6** The 3D UAV trajectory and its Projection for radio surveillance, where the green and orange dots are the UAV initial and final locations, respectively; the red and yellow triangles denote the locations of the Tx and Rx, respectively; and the blue square is the RIS reference point

The average eavesdropping success probability and the average eavesdropping rate are collated under different configurations of the RIS in Table 2.2. It is revealed that under the TD3 algorithm, the successful eavesdropping probability is improved by 22%, and the eavesdropping rate is improved by 26% when $N = 40$, as compared to those when $N = 20$. In addition, under the TD3 algorithm, the eavesdropping probability and rate are improved by 12% and 14%, respectively, as compared to

**Table 2.2** Comparison of eavesdropping capability between the TD3-based algorithms with the baseline schemes, where RDU is based on an ideal assumption with the instantaneous perfect CSI of all channels available to the UAV

| Scheme | Number of RIS elements (N) | Successful eavesdropping probability (%) | Average eavesdropping rate (bit/Hz/s) | Total eavesdropping rate (bit/Hz/s) |
|---|---|---|---|---|
| TD3 | N = 20 | 48.86 | 5.04 | 756.05 |
| | N = 40 | 59.61 | 6.36 | 953.90 |
| RDU | N = 20 | 52.13 | 5.45 | 818.47 |
| DDPG | N = 20 | 43.47 | 4.42 | 663.38 |

the DDPG algorithm. On the other hand, the RDU scheme only enhances the eavesdropping probability and rate by 7% and 8%, respectively, as compared to the TD3 algorithm; see $N = 20$. Although the RDU scheme scores the highest in terms of eavesdropping probability and rate, it is under the ideal assumption of perfect and instantaneous CSI of all channels, which is hardly implementable in practice and would incur a prohibitively high computational complexity.

**Acknowledgements** The heading should be treated as a subsubsection heading and should not be assigned a number.

## References

1. Savkin AV, Huang H (2021) Bioinspired bearing only motion camouflage UAV guidance for covert video surveillance of a moving target. IEEE Syst J 15(4):5379–5382
2. Hu S, Wu Q, Wang X (2021) Energy management and trajectory optimization for UAV-enabled legitimate monitoring systems. IEEE Trans Wirel Commun 20(1):142–155
3. Yuan X, Feng Z, Ni W, Liu RP, Zhang JA, Xu W (2020) Secrecy performance of terrestrial radio links under collaborative aerial eavesdropping. IEEE Trans Inf Forensics Secur 15:604–619
4. Li K, Voicu RC, Kanhere SS, Ni W, Tovard E (2019) Energy efficient legitimate wireless surveillance of UAV communications. IEEE Trans Veh Technol 68(3):2283–2293
5. Hu S, Ni W, Wang X, Jamalipour A, Ta D (2021) Joint optimization of trajectory, propulsion, and thrust powers for covert UAV-on-UAV video tracking and surveillance. IEEE Trans Inf Forensics Secur 16:1959–1972
6. Yang Z, Chen M, Saad W, Xu W, Shikh-Bahaei M, Poor HV, Cui S (2022) Energy-efficient wireless communications with distributed reconfigurable intelligent surfaces. IEEE Trans Wirel Commun 21(1):665–679
7. Yuan X et al (2019) Secrecy rate analysis against aerial eavesdropper. IEEE Trans Commun 67(10):7027–7042
8. Lei H et al (2020) Safeguarding UAV IoT communication systems against randomly located eavesdroppers. IEEE Internet Things J 7(2):1230–1244
9. Hu S, Yuan X, Ni W, Wang X, Jamalipour A (2023) RIS-assisted jamming rejection and path planning for UAV-borne IoT platform: a new deep reinforcement learning framework. IEEE Internet Things J 10(22):20162–20173

10. Duo B, Wu Q, Yuan X, Zhang R (2020) Anti-jamming 3D trajectory design for UAV-enabled wireless sensor networks under probabilistic LoS channel. IEEE Trans Veh Technol 69(12):16288–16293
11. Peng J, Zhang Z, Wu Q, Zhang B (2019) Anti-jamming communications in UAV swarms: a reinforcement learning approach. IEEE Access 7:180532–180543
12. Li Z et al (2021) UAV networks against multiple maneuvering smart jamming with knowledge-based reinforcement learning. IEEE Internet Things J 8(15):12289–12310
13. Yuan X, Hu S, Ni W, Wang X, Jamalipour A (2023) Deep reinforcement learning-driven reconfigurable intelligent surface-assisted radio surveillance with a fixed-wing UAV. IEEE Trans Inf Forensics Secur 18:4546–4560
14. Hu Y, Chen M, Saad W, Poor HV, Cui S (2021) Distributed multi-agent meta learning for trajectory design in wireless drone networks. IEEE J Sel Areas Commun 39(10):3177–3192
15. Hu S, Chen X, Ni W, Hossain E, Wang X (2021) Distributed machine learning for wireless communication networks: techniques, architectures, and applications. IEEE Commun Surv Tutor 23(3):1458–1493. 3rd Quarter
16. Cao Z, Wong K, Lin C-T (2021) Weak human preference supervision for deep reinforcement learning. IEEE Trans Neural Netw Learn Syst 32(12):5369–5378
17. Mu X, Liu Y, Guo L, Lin J, Poor HV (2021) Intelligent reflecting surface enhanced multi-UAV NOMA networks. IEEE J Sel Areas Commun 39(10):3051–3066
18. Sun C, Ni W, Wang X (2021) Joint computation offloading and trajectory planning for UAV-assisted edge computing. IEEE Trans Wirel Commun 20(8):5343–5358
19. You C, Zhang R (2019) 3d trajectory optimization in Rician fading for UAV-enabled data harvesting. IEEE Trans Wirel Commun 18(6):3192–3207
20. Lu H, Zeng Y, Jin S, Zhang R (2021) Aerial intelligent reflecting surface: joint placement and passive beamforming design with 3D beam flattening. IEEE Trans Wirel Commun 20(7):4128–4143
21. Silver D, Lever G, Heess N, Degris T, Wierstra D, Riedmiller M (2014) Deterministic policy gradient algorithms. In: Proceedings of the 31st international conference on international conference on machine learning, series ICML'14, vol 32. JMLR.org, pp I–387–I–395
22. Lillicrap TP, Hunt JJ, Pritzel A, Heess N, Erez T, Tassa Y, Silver D, Wierstra D (2016) Continuous control with deep reinforcement learning. In: Proceedings of the ICLR (poster). [Online]. arXiv:1509.02971
23. Hou Y, Liu L, Wei Q, Xu X, Chen C (2017) A novel DDPG method with prioritized experience replay. In: Proceedings of the IEEE international conference on systems, man and cybernetics (SMC), pp 316–321
24. Sutton RS, McAllester DA, Singh SP, Mansour Y et al (1999) Policy gradient methods for reinforcement learning with function approximation. In: Proceedings NIPS, vol 99. Citeseer, pp 1057–1063
25. Dankwa S, Zheng W (2019) Twin-delayed DDPG: a deep reinforcement learning technique to model a continuous movement of an intelligent robot agent. In: Proceedings of the international conference on vision, image and signal processing, pp 1–5
26. Yuan X, Hu S, Ni W, Liu R-P, Wang X (2023) Joint user, channel, modulation-coding selection, and RIS configuration for jamming resistance in multiuser OFDMA systems. IEEE Trans Commun 71(3):1631–1645
27. Gao G, Li J, Wen Y (2020) DeepComfort: energy-efficient thermal comfort control in buildings via reinforcement learning. IEEE Internet Things J 7(9):8472–8484
28. Mozaffari M, Saad W, Bennis M, Debbah M (2017) Mobile unmanned aerial vehicles (UAVs) for energy-efficient internet of things communications. IEEE Trans Wirel Commun 16(11):7574–7589
29. Cao Y, Lv T, Lin Z, Ni W (2021) Delay-constrained joint power control, user detection and passive beamforming in intelligent reflecting surface-assisted uplink mmwave system. IEEE Trans Cognit Commun Netw 7(2):482–495
30. Fasarakis-Hilliard N, Alevizos PN, Bletsas A (2015) Coherent detection and channel coding for bistatic scatter radio sensor networking. IEEE Trans Commun 63(5):1798–1810

31. Billingsley P (2013) Convergence of probability measures. John Wiley & Sons
32. Redder A, Ramaswamy A, Karl H (2022) Asymptotic convergence of deep multi-agent actor-critic algorithms. arXiv: 2201.00570
33. Fujimoto S, Hoof H, Meger D (2018) Addressing function approximation error in actor-critic methods. In: Proceedings of the international conference on machine learning, pp 1587–1596

# Chapter 3
# Reflection Co-efficient Optimization for Sum-Rate Enhancement in RIS-Assisted UAV Communications

**Maham Misbah and Zeeshan Kaleem**

**Abstract** Unmanned Aerial Vehicles (UAVs) have emerged as a potential solution for beyond 5G networks due to their flexibility and cost-effectiveness. Recently, UAVs have grabbed the attention of researchers due to their wide range of uses, including airborne inspection, drone photography, crop monitoring, delivery services, and wireless communication. In wireless communications, using UAVs in conjunction with ground stations can improve coverage, dependability, and energy efficiency of the network or they can also be deployed as the flying mobile terminals within cellular networks. However, UAVs face several deployment challenges such as optimal placement and height during operation. Another emerging technology, reconfigurable intelligent surface (RIS), is used along with UAVs to further enhance the network efficiency. These low-cost and passive RISs are regarded as the cost-effective alternative for networks beyond 5G and 6G. To broaden their application, they can be deployed on any structure, rooftop, or even UAVs. The RIS-aided UAV networking design has a lot of potential for improving wireless communications and dealing with the increasing complexity of the wireless channel, especially at higher frequencies. The appropriate positioning and elevation of the UAV-RIS are critical for efficient downlink connection between the base station and edge users. Improving the minimum rates of edge users is also an essential factor to consider while implementing this infrastructure. In this chapter, our primary aim is to enhance the overall system rates in an RIS-enhanced UAV network through strategic collaboration with an RIS. The key focus lies in the joint optimization of both UAV altitudes and passive RIS elements. To achieve this, we introduce a novel optimization approach, combining **R**iemannian **G**radient and particle swarm **O**ptimization (**RGO**) techniques. Addressing the inherent complexity of the problem, we adopt an alternating optimization strategy, breaking it down into two interconnected sub-problems. Our

M. Misbah
Department of Electrical and Communication Engineering, United Arab Emirates University (UAEU), Al Ain, United Arab Emirates

Z. Kaleem (✉)
Department of Computer Engineering and the Interdisciplinary Research Center for Smart Mobility and Logistics, King Fahd University of Petroleum and Minerals (KFUPM), Dhahran, Saudi Arabia
e-mail: zeeshan.kaleem@kfupm.edu.sa

© The Author(s), under exclusive license to Springer Nature Singapore Pte Ltd. 2026

I. Ahmad et al. (eds.), *Reconfigurable Intelligent Surfaces for Unmanned Aerial Vehicle Communication*, Unmanned System Technologies,
https://doi.org/10.1007/978-981-96-9662-8_3

experimental results validate the efficiency and superiority of the proposed scheme when compared to baseline strategies.

**Keywords** Particle swarm optimization · RIS phase shifts · Riemannian conjugate gradient · Sum rate · UAV altitude

## 3.1 UAV-Assisted RIS Communications

Unmanned Aerial vehicles (UAVs) or drones have attracted the significant interest of researchers owing to their applicability in vast applications. UAVs can be easily employed to serve the calamity hit area or to further enhance wireless communication. As an alternative to the conventional terrestrial network, UAVs can pave the way for new opportunities such as UAV-aided cellular communication for beyond 5G networks [6]. They can be easily and quickly moved on demand, to increase coverage owing to better line-of-sight (LOS) link with the users at ground, and to have adaptive height to fulfill quality-of-service (QoS) needs dependent on users, required data rate, and interference effects. Furthermore, the mobility of UAVs has unlocked the new areas for improvement by allowing the UAV state to be dynamically adjusted to best fit the communication environment. Adaptive communications and other technologies may also be used with UAV movement control to boost communication performance even further. These clear advantages make UAV-assisted wireless communication an attractive feature of future wireless systems, which will need to serve a wider range of applications while providing orders-of-magnitude capacity improvements over present systems [10]. UAV-assisted relaying, in which UAVs are used to offer wireless connection among different users or user groups who do not have dependable LOS communication links, such as the forefront and the control center for emergency response. Figure 3.1 shows the different use cases of UAVs.

Despite all these applications, UAVs are subjected to some challenges as well. Unlike the ground base stations that are fixed and path loss only depends upon the user location. In aerial base stations (ABS), the location of the user, and UAV contributes to the path loss. Deployment of the aerial base station or UAV poses a significant challenge when it comes to enhancing network efficiency. Another challenge faced by UAVs is the limited transmit power and different requirement for signal-to-noise power ratio (SNR). Reconfigurable intelligent surface (RIS), an emerging technology often named as intelligent reflecting surface or software controlled surface, is a two-dimensional meta-surface comprising of a large number of reflecting elements whose phase and amplitude can be adjusted in the wireless environment for directed communication and better channel controls [18]. Their adaptive nature to intelligently control the received signals helps to attain many purposes such as signal enhancement at the user side or reducing the effect of interference. The passive nature of RIS cuts down the energy consumption thus solving the major problem of next-generation networks. Future cellular network deployments will benefit greatly from ABS and Aerial Relay (AR) technology, which are viewed as essential enablers for dynamic

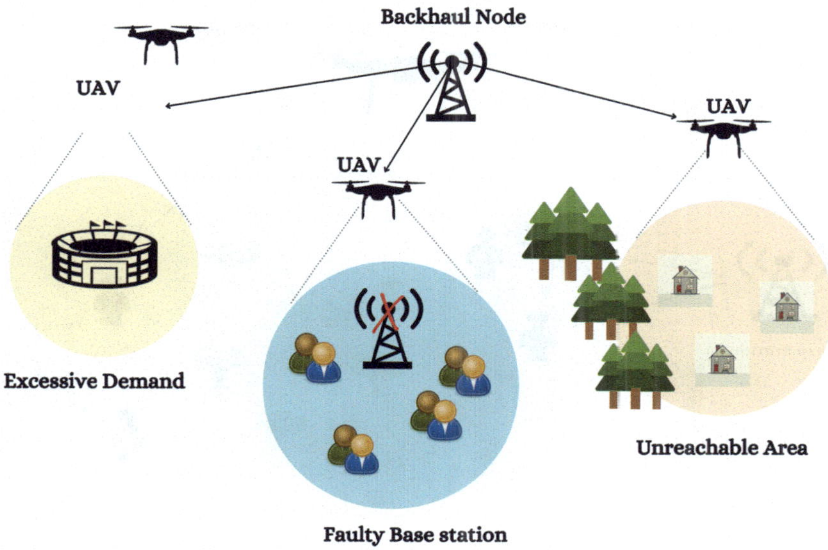


**Fig. 3.1** Different use cases of UAVs

and adaptable coverage. However, these active aerial communications need a large amount of energy [26]. The RIS is a more passive option than ABS and AR. Moreover, AR utilizes the half-duplex (HD) mode, which is less efficient compared to full-duplex (FD). Another key advantage over active FD relay systems is that the RIS's relaying method is resistant to the amplification and self-interference caused by antenna noise. Figure 3.2 illustrates a scenario of the multiple users where the base station is unable to provide complete coverage due to hindrance and long distance. A UAV-assisted RIS is employed in that area to provide better coverage by beam-forming signal with the help of RIS. Moreover, RIS also helps in the cancellation of unwanted signals and interference from the nearby base stations.

### *3.1.1 Challenges in UAV-RIS Integration*

RIS-aided UAV architecture poses significant challenges when it comes to their deployment such as channel modeling, RIS phase adjustment, and UAV-RIS positioning. A lot of considerations are required for developing RIS-UAV channels [2]. The two components of this architecture, RIS and UAV, make the channel modeling more complex. Mobility of the UAV and shadowing effects leads to varying environmental conditions and RIS further adds to this complexity due to its passive nature of elements. The passive nature of the RIS, on one hand, provides a cost-effective and energy-efficient solution but, on the other hand, this passive nature makes channel

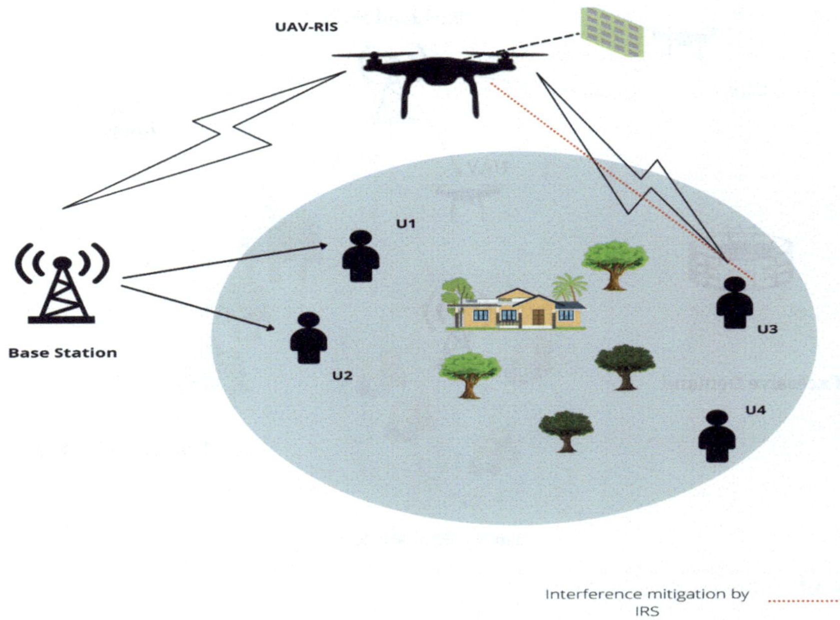


**Fig. 3.2** Signal enhancement and interference mitigation by UAV-RIS for edge users

estimation more difficult, therefore some low-power active elements are required to tackle this situation. A fully controlled and synchronized link is required between RIS and its controller which is subjected to continuous variations. Solving these challenges can open new ways to implement RIS-aided UAVs seamlessly in next-generation wireless networks by increasing reliability and efficiency. The current research aims at maximizing the minimum sum rate of the users in RIS-assisted UAV networks by taking into account the UAV height, transmit power, and RIS angle variations.

### *3.1.2 Enhancing UAV-RIS Efficiency*

The effectiveness and efficiency of any system decide its practicality and performance in real-world cases. UAV's and RIS's optimal heights and locations, RIS phases, and interference have all been taken into account while developing UAV-assisted RIS system. By optimizing beam-forming vectors of RIS, either the ground station or RIS or drone notably increases the system's efficiency by maximizing system sum rate [17]. Therefore, studying the impact of these parameters in UAV-assisted RIS in various situations provides useful insights for developing future networks [8].

### *3.1.3 Comparison of UAV Air-to-Ground Propagation Model with Terrestrial Model*

The terrestrial communication channels, such as the urban channel, have traits that the Air-to-Ground (A2G) channel does not have. On one hand, satellite communications have the obvious advantage over terrestrial ones in that LOS propagation. It is possible to reduce diffraction and shadow losses by using an A2G channel in situations when only non-LOS (NLOS) pathways are available and the angle to the UAV is significant. UAV velocities allow the A2G channel to display substantially larger rates than traditional terrestrial channels. Whenever the channel is simulated statistically, it might indicate that the channel's statistics are almost constant. It's possible that the channel's characteristics will change extremely slowly if the UAV is not near scattering elements or the Ground Station (GS). This is particularly true for hovering UAVs. There may be delay of time when the received signal is completely lost, thus typical communication strategies like interleaving and averaging are ineffective [12].

### *3.1.4 UAV A2G Propagation Frequency Bands and Channel Specifications*

Frequency band is a critical component in all channels of communication due to its effect on propagation characteristics. With regard to the unlicensed Control and Non-Payload Communication (CNPC) payload bands and existing unlicensed bands, tropospheric attenuation due to atmospheric gases or hydrometers is minimal. At higher frequencies, such as Ku, Ka, and other mm-wave (mmWave) bands, such as 100 GHz or more, this will not be the case. As a result, the free-space path loss (PL) and tropospheric attenuation in these frequency range may be greater. Because of this, short-range A2G connections are often deployed in these frequencies [7].

Due to refraction from the ground, terrestrial structures (ground scatterers), and even the UAV itself, multipath components (MPCs) occur in an A2G propagation channel. The composition, geometry, and size of the dispersing items will influence the channel's properties. Single reflection off the earth's surface is frequently the strongest MPC separated from the LoS part in an A2G propagating scenario resulting in a two-way model [24].

### *3.1.5 RIS Channel Modeling*

Path loss modeling for RIS-assisted communication networks has received a great deal of attention in the past, but simulating the end-to-end channels has also lately received considerable interest. When estimating the overall path loss of RIS-assisted systems, the impacts of RISs in the near and distant fields should be taken into account. The approach used to represent the end-to-end channel is also crucial [25]. Researchers have pursued both physics-based and statistical channel modeling methodologies, focusing on the statistical qualities of scatterers and groups in the environment while analyzing the physical aspects of reflections and scattering effects. As an added bonus, there are a number of electromagnetic (EM)-compatible channel models that take into consideration EM features such as impedance matching among the RIS's sub-wavelength unit cells. Non-invasively enriching the surroundings of current wireless networks for better spectrum efficient energy conservation without changing wireless protocols and architectures for transceivers currently in deployment is made possible by the passive nature of an RIS.

### *3.1.6 Amplitude and Phase Shift Control*

The digitally controlled meta-surface is a reflecting surface filled with PIN diodes. The reflection factors of the surface components may be dynamically adjusted by supplying controlling voltages to a specific PIN diode. A greater degree of control over the variables allows for more design flexibility. Thus, improved results are possible. However, one of the disadvantages of falling into this group is the substantial rise in complexity.

It is possible to adjust the phase shift of the reflected signal in a passive RIS by using a great amount of passive components. It is common for each passive RIS element to include an impedance-adjustable circuit to alter the phase of the signal. Active radio-frequency (RF) elements are required to operate a passive RIS element, thus it uses almost little direct current power and the thermal noise it introduces is generally insignificant [21]. The formulation and solution of optimization issues are common in RISs built using the phase shift. Modeling errors might make a solution less than optimum. As long as the exact phase shift can't be predicted, an error model for the phase shift must be used to assess the design algorithms' performance sensitivity in light of the modeling parameters.

## 3.2 System Sum-Rate Maximization

The UAV can offer greater coverage in the event of barriers or high-rise structures; this may be further improved by deploying many UAVs to meet the needs of extended coverage. By meeting quality-of-service (QoS) limitations, the coverage area may be expanded and users' mean and edge rate can be enhanced by deploying numerous UAVs. The joint optimization of various parameters like ABS altitude, placement, power allocation, and trajectory will further improve the link. RIS has drawn interest recently as a means of controlling communication quality and the propagation environment [3]. In order to align the signals from various pathways to the intended receiver and enhance the signal quality and sum rate, RIS is made up of several movable passive reflecting components with their phase shifts collaboratively controlled [5]. An innovative ABS 3D positioning that optimizes the number of covered users with varying quality-of-service (QoS) requirements is presented in [22]. They proposed an exhaustive search (ES) scheme to optimize the ABS placement. In [1], the authors proposed the solution to increase a cellular system's downlink data throughput by efficiently deploying UAVs. A reinforcement-learning aided method was proposed to determine the optimal location of a UAV to assist a macro-base station. The UAV location was optimized to enhance the overall data rates of all users. The combination of UAV and RIS has given us an innovative research direction for improvements. The benefits of RIS-assisted UAV communications can be further enhanced by optimizing the RIS phase shift and UAV placement [20]. Several studies have pointed toward this direction. In [14], the authors deployed an RIS-assisted UAV system by using non-orthogonal multiple access to enhance the spectral efficiency of wireless networks. Here, the authors formulated the energy consumption minimizing problem for jointly optimizing RIS phase shifts, power allocation, and user's mobility and solved it using a decaying deep Q-network algorithm. Link performance and coverage problems of aerial-ground communication were discussed in [4]. They modeled the RIS-aided UAV transmission strategy as a mixed non-integer non-linear programming optimization problem to maximize the overall system throughput. Moreover, the researchers employed multitask learning for fast computation. In [9], a new RIS-assisted communication model was proposed that investigates the UAV trajectory and RIS phase shift's impacts on the system's sum rate and optimizes them to maximize the data rate. An alternating optimization technique was adopted, which outperforms the benchmark schemes. For instance, the similar and complex issue of RIS-enhanced UAV network was solved by the authors in [23] using an alternating optimization technique. By carefully adjusting the UAV 3D path, balanced resource distribution, and RIS phase shifters, they were able to overcome the issue. Furthermore, the writers also take the QoS requirements of the consumers seriously. In [15], UAV 3D space and RIS phase shifts were optimized using double deep Q-network and deep deterministic policy gradient. They aimed at maximizing the UAV data transfer while minimizing the energy propulsion for RIS-assisted UAV systems. Similarly, in [16], the joint optimization of UAV trajectory and cache with the phase shift design of the RIS was presented. They adopted the successive convex approximation scheme to solve the problem and verified the results by getting better performance than the benchmark schemes.

The attainable rate was maximized in [13] by simultaneously adjusting the RIS phase shifts and the UAV trajectory. The non-convex challenge was broken down into smaller problems by optimizing the UAV trajectory and passive beam-forming. In order to align the received signals, they first adjusted the phase shifts of the signals for a certain UAV trajectory. The sub-optimal UAV trajectory was then obtained by using the successive convex approximation with the optimal phase.

For enhancing system sum rate, RIS-assisted UAV network is considered as shown in Fig. 3.3. RIS comprising of $\mathbb{N}$ reflecting elements is mounted on the building to aid the low-altitude UAVs by controlled reflection toward the receiver. The ABS, user, and RIS elements are denoted by $\mathcal{U} \triangleq 1, 2, \ldots, u$, $\mathcal{D} \triangleq 1, 2, \ldots, d$, and $\mathcal{R} \triangleq$

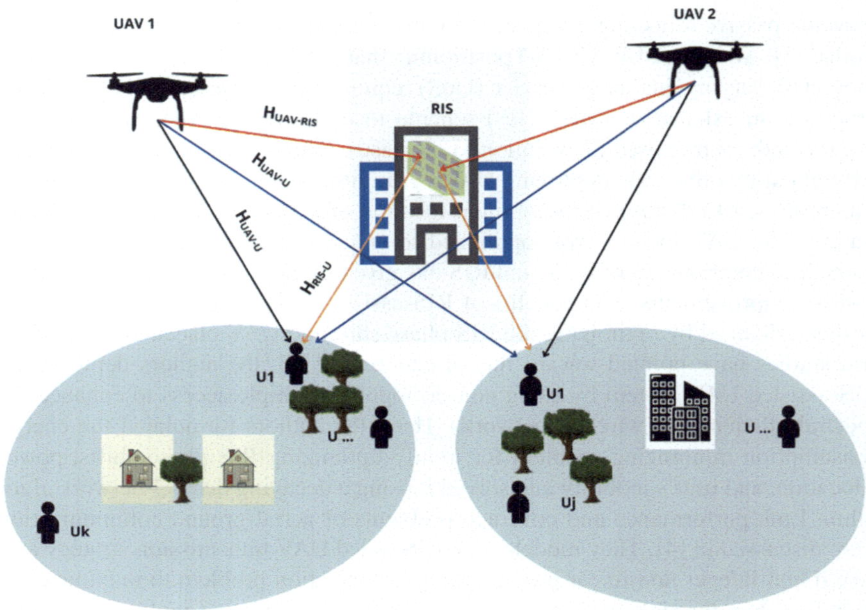


**Fig. 3.3** System model

$1, 2, \ldots, r$, respectively. The 3D positions of UAV, RIS, and user are denoted as $\mathbf{p}u = [\mathbf{q}u, h_u]$, $\mathbf{p}r = [\mathbf{q}r, h_r]^T$, $\mathbf{p}d = [\mathbf{q}d, 0]^T$ with $\mathbf{q} = (x, y)$. The reflection co-efficient $\boldsymbol{\Phi}$ with $\mathbb{N}$ number of reflecting elements is defined as $\text{diag}(\alpha_1 e^{j\phi_1}, \ldots, \alpha_N e^{j\phi_N})$ where $\phi \in [0, 2\pi)$ and $\alpha_n = \alpha_1, \ldots, \alpha_N \in [0, 1]$ are the reflection and amplitude coefficients, respectively.

The distance between the UAV and RIS is given as $D_{u,r} = \sqrt{|\mathbf{p}r - \mathbf{p}u|^2}$, between user and RIS is given by $D_{r,d} = \sqrt{|\mathbf{p}d - \mathbf{p}r|^2}$, and between user and UAV is given by $D_{d,u} = \sqrt{|\mathbf{p}u - \mathbf{p}d|^2}$.

The channel co-efficient between $u - th$ UAV and $d - th$ user is modeled as $h_{ud} = \sqrt{L_{ud}} g_{ud}$. The $g_{ud}$ here is modeled as Rician fading:

$$g_{ud} = \sqrt{\frac{K_c}{K_c + 1}} \bar{g} + \sqrt{\frac{1}{K_c + 1}} \tilde{g}$$

Here, $K_c$ is the Rician factor [11] and the LoS and NLoS components are given by $\bar{g}$ and $\tilde{g}$, respectively.

The channel model $\mathbf{h}_{ur}$ between UAV and RIS is modeled as LoS channel given by

$$\mathbf{h}_{ur} = \sqrt{l_{ur}} \sum_{n=1}^{N} \alpha_n \left| \boldsymbol{g}_{ur_n} \right| e^{j(\phi_n - \varphi_n)}$$

Similarly, the channel model $\mathbf{h}_{rd}$ between user and RIS is given as

$$\mathbf{h}_{rd} = \sqrt{l_{rd}} \sum_{n=1}^{N} \alpha_n \left| g_{rd_n} \right| e^{j(\phi_n - \theta_n)}$$

The distance-based channel gain $L_{ij}$ (dB) is calculated as

$$L_{ij} = \frac{I_1}{1 + ae^{(-b\frac{180}{\pi}\zeta_e - a)}} + 20 \log \left( \frac{d_{i,j}}{\cos \zeta} \right) + I_2$$

Here, $I_1 = \eta_{LoS} - \eta_{NLoS}$, $I_2 = 20 \log \frac{(4\pi f_c d_{i,j})}{c} + \eta_{LoS}$, and $q_i^2 + d_{i,j}^2 = \left( \frac{d_{i,j}}{\cos \zeta} \right)^2$. In linear scale the path loss is modeled as $l_{ij} = 10^{\frac{L_{ij}}{10}}$.

Based on the signal-to-interference plus noise ratio (SINR), the system sum rate is calculated as

$$C = \log_2 \left( 1 + \frac{p_{T_d} |(\mathbf{h}_{ur} \boldsymbol{\Phi} \mathbf{h}_{rd} + h_{ud}|^2}{\sum_{d=1, i \neq d} p_{T_i} |(\mathbf{h}_{ur} \boldsymbol{\Phi} \mathbf{h}_{ri} + h_{ui}|^2 + \sigma^2} \right)$$

where $p_T$ is the UAV transmit power and $\sigma^2$ is the Gaussian distribution noise signal.

## 3.3 Sum-Rate Maximization Scheme

The sum-rate maximization problem for varying UAV transmit power, UAV heights, and RIS phase shifts is formulated as follows:

$$\max_{\mathbf{q}_u, \phi, p_t} \sum_{u=1}^{U} \sum_{d=1}^{D} C_{u,d}, \forall u \in \mathcal{U} \quad (3.1a)$$

$$[x_{\min}, y_{\min}, q_{\min}] \leq (\mathbf{q}_u, h_u) \leq [x_{\max}, y_{\max}, q_{\max}] \quad (3.1b)$$

$$p_{T_u} \geq 0 \quad (3.1c)$$

$$\phi_n \in [0, 2\pi), \quad \forall n \in N \quad (3.1d)$$

$$\alpha_n \in [0, 1). \quad (3.1e)$$

Constraint (3.1b) gives the UAV 3D positioning region. Constraint (3.1b) keeps track of the UAV transmit power $p_{T_u}$ to be greater than zero. Constraints (3.1d) and (3.1e) keep the RIS phase and amplitude within specified range.

For solving this problem, we propose **R**iemannian **G**radient **O**ptimization (**RGO**) scheme. The above-mentioned problem is difficult to solve due to its non-convex nature. Therefore, for the ease of implementation we have sub-divided the problem into two further parts. Initially, our objective is to calculate the optimal RIS phase shift. We employ the Riemannian conjugate gradient method, accounting for variations in UAV transmit power while maintaining a fixed altitude. This ensures the best RIS configuration under different operational conditions. Subsequently, we extend our optimization efforts to UAV altitude. Utilizing the optimized RIS phase shifts obtained in the first step, we employ particle swarm optimization to determine the UAV height that maximizes the overall system sum rate.

For the ease of computation, we first optimize the 2D UAV positions $\mathbf{P}^* = (x_u, y_u)$ using $K$-means algorithm [19]. The $K$-means algorithm is used for optimizing the 2D position of UAV in wireless communication networks with multiple users. By dividing the users into clusters and associating each cluster with a UAV, the $K$-means algorithm aims to find a feasible solution of UAV positions to maximize the total system sum rate by satisfying the constraints. The $K$-means algorithm achieves this by iteratively clustering the users based on their location and updating the UAV positions to minimize the distance between users and their associated UAVs. An optimal UAV height is then obtained using $\phi_s^*$ and $\mathbf{P}^*$. This two-step optimization process allows us to navigate the complex interplay between UAV altitudes and RIS configurations, ultimately improving the network's performance in terms of system sum rate.

### 3.3.1 *Optimizing RIS Phase Shift Coefficients*

RIS phase shift optimization is crucial for directed beam-forming design, given the UAV height. Let $\boldsymbol{h}_{vn} = \text{diag}(\boldsymbol{h}_{ur})\boldsymbol{h}_{rd}$, as $\boldsymbol{\phi}\boldsymbol{h}_{ur}\boldsymbol{h}_{rd} = \boldsymbol{\theta}\,\text{diag}(\boldsymbol{h}_{ur})\boldsymbol{h}_{rd}$. Consequently, the SINR for the $d-th$ user is modified to $\gamma_d^{'}$, and the modified sum rate is calculated as $C(\boldsymbol{\vartheta}) = \log_2\left(1 + \frac{p_{T_d}|(\boldsymbol{h}_{ur}\boldsymbol{\vartheta}\boldsymbol{h}_{vd}+h_{uk}|^2}{\sum_{d=1,i\neq k}^{K} p_{T_i}|(\boldsymbol{h}_{ur}\boldsymbol{\vartheta}\boldsymbol{h}_{vi}+h_{ui}|^2+\sigma^2}\right)$.

Given the varying UAV transmit power $p_T$ and given UAV height $\mathbf{P}_u$ for phase shift optimization using the RIS is re-modeled as

$$\max_{\boldsymbol{\vartheta}, p_T} C(\boldsymbol{\vartheta}) \tag{3.2a}$$

$$\text{s.t. } (1c), (1d), (1e). \tag{3.2b}$$

The objective function described above is continuous and differentiable, which can converge quickly using Riemannian algorithm [27]. Steps involved are outlined as follows:

- First the Riemannian conjugate gradient($\mathsf{g}$) is calculated, which is denoted as the projection of Euclidean gradient ($\nabla C$) as
  $\mathsf{g} = \nabla C(\boldsymbol{\vartheta}) - \text{Re}\left\{\nabla C(\boldsymbol{\vartheta}) \circ \boldsymbol{\vartheta}^*\right\} \circ \boldsymbol{\vartheta}$. The $\nabla C$ is determined as

$$\nabla C(\boldsymbol{\vartheta}) = \sum_{d=1}^{D} 2p_T \left( \frac{\sum_i \boldsymbol{h}_{vi}\boldsymbol{\vartheta} + \sum_i \boldsymbol{h}_{vi}h_{ui}}{\sum_i p_T \left|\boldsymbol{\vartheta}\boldsymbol{h}_{vi} + h_{ui}\right|^2 + \sigma^2} - \frac{\sum_{i\neq d} \boldsymbol{h}_{vi}\boldsymbol{\vartheta} + \sum_{i\neq k} \boldsymbol{h}_{vi}h_{ui}}{\sum_{i\neq d} p_t \left|\boldsymbol{\vartheta}\boldsymbol{h}_{vi} + h_{ui}\right|^2 + \sigma^2} \right). \tag{3.3}$$

- In the next step, conjugate tangent to $\mathsf{g}$, also known as search direction $\mathsf{d}$, is calculated as $\mathsf{d}_s = -\mathsf{g} + \varrho\chi\,(\mathsf{d}_{s-1})$, here $\varrho_1$ is the updating gradient parameter and $\chi\,(\mathsf{d}_{s-1}) = \mathsf{d}_{s-1} - \mathrm{Re}\{\mathsf{d}_s \circ \theta^*\}\vartheta$, with preceding search direction $\mathsf{d}_{s-1}$ and $\chi\,(\mathsf{d}_{s-1})$ is the vector transfer function.
- The armijo step size $\varrho_2$ is updated with each iteration to get the optimized value of $\vartheta_s$ as $\boldsymbol{\vartheta}_s \leftarrow \frac{(\boldsymbol{\vartheta}+\varrho_2\mathsf{d}_s)}{|(\boldsymbol{\vartheta}+\varrho_2\mathsf{d}_s)|}$, and iteration accuracy ($\delta$) is set to $10^{-6}$ [27].

---

**Algorithm 3.1:** Conjugate Gradient Algorithm for RIS phase shift optimization

---

1 **Input:** $\boldsymbol{\vartheta}, p_T, \sigma, \varrho_1, \varrho_2, \delta$
2 **Output:** $\boldsymbol{\vartheta}_s^*$ **Data**: Initialize: $\boldsymbol{\vartheta}_0$ and iteration $s = 0$
3 **while** $|\mathsf{g}_s| > \delta$ **do**
4 Initial Gradient using

$$\mathsf{g}_s = \nabla C(\boldsymbol{\vartheta}_s) - \mathrm{Re}\left\{\nabla C(\boldsymbol{\vartheta}_s) \circ \boldsymbol{\vartheta}_s^*\right\} \circ \boldsymbol{\vartheta}_s$$

Calculating search direction

$$\mathsf{d}_s = -\mathsf{g}_s + \eta_1\tau\,(\mathsf{d}_{s-1})$$

get $\boldsymbol{\vartheta}_s$ based on $\boldsymbol{\vartheta}_s \leftarrow \frac{(\boldsymbol{\vartheta}_s+\eta_2\mathsf{d}_s)}{|(\boldsymbol{\vartheta}_s+\eta_2\mathsf{d}_s)|}$ get $s = s + 1$

5 **Output:** $\boldsymbol{\vartheta}_s^*$against each value of $p_T$ for given $\boldsymbol{\vartheta}$

---

### 3.3.2 *Optimizing UAV Heights Using PSO*

The given sum-rate maximization problem results in the sub-optimal solution and heuristic algorithms such as PSO are best suited for such problems. The UAV heights are optimized here using PSO considering fixed UAV transmit power, 2D UAV positions $\boldsymbol{P}^*$, and optimal RIS phase shifts $\boldsymbol{\vartheta}_s^*$. The sub-problem is presented as

$$\max_{q_u} \sum_{u=1}^{U}\sum_{d=1}^{D} C_{u,d} \tag{3.4a}$$

$$\text{s.t. } (1b), (1c). \tag{3.4b}$$

Here in (3.1c), the UAV height is optimized by maximizing the objective function given in (3.4a). The PSO, a meta-heuristic algorithm, iteratively optimizes the UAV heights which is based on global and local best.

In the initial step, we begin with the creation of a particle set ($pop$) in which each particle is assigned random positions representing the altitudes of the assisted aerial base stations (ABS) ($q_{ui}$). The particle set also initializes with zero velocities ($\boldsymbol{\Omega}_i$), where $i$ denotes the index of the particle. The utility function employed for optimization is the system sum rate ($C$), and the position of the particle with the best utility is denoted as $q_{uG}$.

Subsequently, in each iteration, particle velocities are updated using the information from the global best position ($q_{uG}$). This update is governed by the following formula:

$$\Omega_i^{(j+1)} = \varepsilon \Omega_i^{(j)} + c_1 \varepsilon_1 \left( q_{ui,\ell}^{(j)} - q_{ui}^{(j)} \right) + c_2 \varepsilon_2 \left( q_{uG}^{(j)} - q_{ui}^{(j)} \right)$$

Here, $j$, $\varepsilon_1$, and $\varepsilon_2$ represent the iteration index, local learning co-efficient, and global learning co-efficient, respectively. Each particle then updates its position based on its velocity using the following formula:

$$q_{ui}^{(j+1)} = \Omega_i^{(j+1)} + q_{ui}^{(j)}$$

The algorithm keeps track of optimal positions for each particle based on their utility values. This process continues iteratively until reaching the maximum iteration limit $j_{\max}$. Finally, the algorithm yields the best solution corresponding to the global best position found in the last iteration. Algorithm 3.2 provides a detailed outline of the altitude optimization process using Particle Swarm Optimization (PSO).

**Algorithm 3.2:** Particle Swarm Optimization for UAV height optimization

**Input:** Initial particle positions and velocities, maximum iteration $j_{\max}$, learning coefficients $\varepsilon, \varepsilon_1, \varepsilon_2$
**Output:** Optimal particle positions $q_{ui}^*$
**Initialize:** $pop, \boldsymbol{\Omega}_i, q_{uG}, C, j = 0$
**repeat:** Update particle velocities using $q_{uG}$:

$$\Omega_i^{(j+1)} = \varepsilon_i^{(j)} + c_1 \varepsilon_1 \left( q_{ui,\ell}^{(j)} / - q_{ui}^{(j)} \right) + c_2 \varepsilon_2 \left( q_{uG}^{(j)} / - q_{ui}^{(j)} \right)$$

Update particle positions:

$$q_{ui}^{(j+1)} = \Omega_i^{(j+1)} + q_{ui}^{(j)}$$

Update $C$ and $q_{uG}$ based on the best utility
Update $j = j + 1$
**Until** $j \leq j_{\max}$
**Output:** $q_{ui}^*$ for each particle

## 3.4 Results and Discussions

To prove the efficacy of the proposed RGO scheme for joint UAV height and RIS phase shift optimization we consider two scenarios for comparison, without RIS and using HPPO, where RIS phases are determined using proximal policy gradient. In RIS-assisted UAV network, 2 UAVs are serving 25 users in two different coverage areas with $500m$ diameter with the help of RIS. The UAV is horizontally positioned at (−40m, 80m) and (40m, 58m). RIS is set at the height of 45 m having 2D co-ordinates of (0, 40m).

The parameters for PSO algorithms are set as, population size, $pop = 170$ having step size, $\Delta = 20$ and $(\Omega_1, \Omega_2) = (2, 2)$ are the learning coefficients. The minimum and maximum transmit power of UAV is set as $[p_{\min}, p_{\max}] = [0, 18]$dBm and the noise power is given by $\sigma^2 = -90$dB.

To prove the superiority of our scheme, we implemented and compared two different scenarios, without RIS and with RIS phase estimated through proximal policy (HPPO). We plot the system sum rate by varying UAV heights from 20 m to 160 m with constant transmit power (20dBm). The highest sum rate of 32bps/Hz is achieved at the height of 40 m using the proposed RGO scheme. The highest rates achieved for the same UAV height without RIS scheme and HPPO are 26.9bps/Hz and 29.8bps/Hz, respectively. The sum rate shows a decreasing trend with increased UAV height because of the increased interference and path loss. In the HPPO scheme, the inter-cell interference was completely ignored, which resulted in deteriorated performance.

In Fig. 3.4b, varying UAV transmit power is plotted against the sum rate for proposed RGO, HPPO, and without RIS setup. An increasing trend for system sum rate is demonstrated by all three schemes with the increase in UAV transmit power $p_T$. However, the proposed RGO scheme showed enhanced performance by achieving a high sum rate at low $p_T$ compared to the HPPO and without the RIS scheme because it takes into account the near cell interference. The without RIS scheme is the least efficient due to the unavailability of the LoS path, and the HPPO also shows 6–7% less efficiency compared to the proposed RGO scheme.

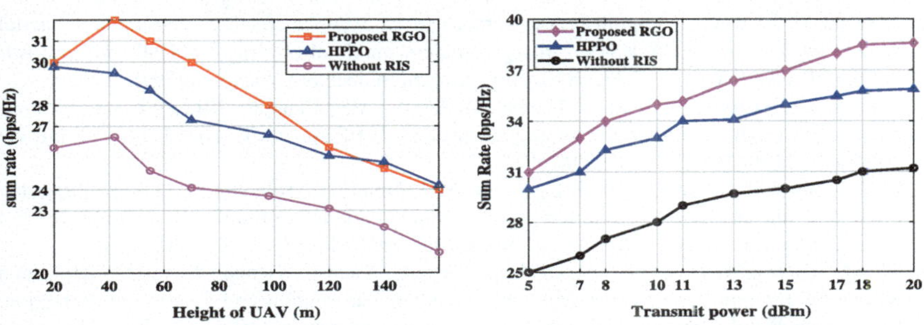


(a) Varying UAV heights (b) Varying UAV transmit power $p_T$

**Fig. 3.4** Impact on system sum rate by varying **a** UAV heights, **b** Transmit power $P_T$

## 3.5 Conclusion

UAVs are flexible and cost-effective alternative for beyond 5G networks, owing to their effectiveness in terms of flexibility, coverage, and reliability in different scenarios. Despite all these benefits, UAV-based wireless networks are subjected to many problems such as path loss and NLoS linkages, particularly in case of low-flying UAVs. Another emerging cost-effective technology, RIS, is being considered for beyond 5G and 6G networks due to their passive nature and enhanced interference mitigation capabilities. Moreover, they can be mounted on any building, rooftop, or UAV. The RIS-aided UAV networking design might improve wireless communications and handle the rising complexity of the wireless channel, particularly at higher frequencies. Positioning and height of the UAV-RIS are crucial for achieving higher system sum rates in wireless communication systems. Current work deploys RIS-assisted UAVs to improve the system sum rates using proposed RGO scheme. We maximized the system sum rate by considering UAV height, transmit power, RIS phase shifts, and inter-cell interference. UAV heights are optimized for better coverage using PSO algorithm and conjugate gradient further optimizes the RIS phase shifters for better LoS linkage. This joint optimization resulted in 6.2% higher sum rates compared to the state-of-the-art HPPO scheme.

## References

1. Al-Ahmed SA, Shakir MZ, Zaidi SAR, Optimal 3D UAV base station placement by considering autonomous coverage hole detection, wireless backhaul and user demand. 22(6):467–475. https://doi.org/10.23919/JCN.2020.000034. https://ieeexplore.ieee.org/document/9321160/
2. Alzenad M, El-Keyi A, Lagum F, Yanikomeroglu H, 3-D placement of an unmanned aerial vehicle base station (UAV-BS) for energy-efficient maximal coverage. 6(4):434–43. https://doi.org/10.1109/LWC.2017.2700840. http://ieeexplore.ieee.org/document/7918510/
3. Basar E, Di Renzo M, De Rosny J, Debbah M, Alouini MS, Zhang R (2019) Wireless communications through reconfigurable intelligent surfaces. IEEE Access 7:116753–116773. https://doi.org/10.1109/ACCESS.2019.2935192
4. Cao X, Yang B, Huang C, Yuen C, Renzo MD, Niyato D, Han Z (2021) Reconfigurable intelligent surface-assisted aerial-terrestrial communications via multi-task learning. IEEE J Sel Areas Commun 39(10):3035–3050. https://doi.org/10.1109/jsac.2021.3088634
5. Cheng Y, Peng W, Huang C, Alexandropoulos GC, Yuen C, Debbah M (2022) RIS-aided wireless communications: extra degrees of freedom via rotation and location optimization. IEEE Trans Wirel Commun 21(8):6656–6671. https://doi.org/10.1109/TWC.2022.3151702
6. Cicek CT, Gultekin H, Tavli B, Yanikomeroglu H, UAV base station location optimization for next generation wireless networks: overview and future research directions. In: 2019 1st international conference on unmanned vehicle systems-oman (UVS). IEEE, pp 1–6. https://doi.org/10.1109/UVS.2019.8658363. https://ieeexplore.ieee.org/document/8658363/
7. Cui Z, Briso C, Guan K, Matolak DW, Calvo-Ramírez C, Ai B, Zhong Z, Low-altitude UAV air-ground propagation channel measurement and analysis in a suburban environment at 3.9 GHz. 13(9):1503–150. https://doi.org/10.1049/iet-map.2019.0067. https://onlinelibrary.wiley.com/doi/10.1049/iet-map.2019.0067
8. Farrag S, Maher EA, El-Mahdy A, Dressler F (2023) Sum rate maximization of uplink active RIS and UAV-assisted THz mobile communications. In: 2023 19th international conference

on the design of reliable communication networks (DRCN), pp 1–7. https://doi.org/10.1109/DRCN57075.2023.10108256
9. Ge L, Zhang H, Wang JB, Joint placement and beamforming design in multi-UAV-IRS assisted multiuser communication. In: 2021 IEEE global communications conference (GLOBECOM). IEEE, pp 01–06. https://doi.org/10.1109/GLOBECOM46510.2021.9685505. https://ieeexplore.ieee.org/document/9685505/
10. Hashida H, Kawamoto Y, Kato N, Intelligent reflecting surface placement optimization in air-ground communication networks toward 6G. 27(6):146–15. https://doi.org/10.1109/MWC.001.2000142. https://ieeexplore.ieee.org/document/9183763/
11. Hua M, Yang L, Wu Q, Pan C, Li C, Swindlehurst AL (2021) UAV-assisted intelligent reflecting surface symbiotic radio system. IEEE Trans Wirel Commun 20(9) 5769–5785
12. Khawaja W, Ozdemir O, Erden F, Guvenc I, Matolak DW, Ultra-wideband air-to-ground propagation channel characterization in an open area. 56(6):4533–455. https://doi.org/10.1109/TAES.2020.3003104. https://ieeexplore.ieee.org/document/9119755/
13. Li S, Duo B, Yuan X, Liang YC, Di Renzo M (2020) Reconfigurable intelligent surface assisted UAV communication: joint trajectory design and passive beamforming. IEEE Wirel Commun 9(5):716–720. https://doi.org/10.1109/lwc.2020.2966705
14. Liu X, Liu Y, Chen Y (2021) Machine learning empowered trajectory and passive beamforming design in UAV-RIS wireless networks. IEEE J Sel Areas Commun 39(7):2042–2055. https://doi.org/10.1109/jsac.2020.3041401
15. Mei H, Yang K, Liu Q, Wang K (2022) 3D-trajectory and phase-shift design for RIS-assisted UAV systems using deep reinforcement learning. IEEE Trans Veh Technol 71(3):3020–3029. https://doi.org/10.1109/TVT.2022.3143839
16. Mei H, Yang K, Shen J, Liu Q (2021) Joint trajectory-task-cache optimization with phase-shift design of RIS-assisted UAV for MEC. IEEE Wirel Commun Lett 10(7):1586–1590. https://doi.org/10.1109/LWC.2021.3074990
17. Misbah M, Kaleem Z, Khalid W, Yuen C, Jamalipour A (2023) Phase and 3-D placement optimization for rate enhancement in RIS-assisted UAV networks. IEEE Wirel Commun Lett 12(7):1135–1138. https://doi.org/10.1109/LWC.2023.3263224
18. Nguyen KK, Khosravirad S, da Costa DB, Nguyen LD, Duong TQ, Reconfigurable intelligent surface-assisted multi-UAV networks: efficient resource allocation with deep reinforcement learning. arXiv:2105.14142
19. Nouri N, Abouei J, Sepasian AR, Jaseemuddin M, Anpalagan A, Plataniotis KN (2022) Three-dimensional multi-UAV placement and resource allocation for energy-efficient IoT communication. IEEE Internet Things J 9(3):2134–2152. https://doi.org/10.1109/jiot.2021.3091166
20. Pogaku AC, Do DT, Lee BM, Nguyen ND (2022) UAV-assisted RIS for future wireless communications: a survey on optimization and performance analysis. IEEE Access 10:16320–16336. https://doi.org/10.1109/access.2022.3149054
21. Praia J, Pavia JP, Souto N, Ribeiro M, Phase shift optimization algorithm for achievable rate maximization in reconfigurable intelligent surface-assisted THz communications. 11(1):18. https://doi.org/10.3390/electronics11010018. https://www.mdpi.com/2079-9292/11/1/18
22. Shakhatreh H, Khreishah A, Alsarhan A, Khalil I, Sawalmeh A, Othman NS, Efficient 3D placement of a UAV using particle swarm optimization. In: 2017 8th international conference on information and communication systems (ICICS). IEEE, pp 258–263. https://doi.org/10.1109/IACS.2017.7921981. http://ieeexplore.ieee.org/document/7921981/
23. Wei Z, Cai Y, Sun Z, Ng DWK, Yuan J, Zhou M, Sun L (2021) Sum-rate maximization for IRS-assisted UAV OFDMA communication systems. IEEE Trans Wirel Commun 20(4):2530–2550. https://doi.org/10.1109/TWC.2020.3042977
24. Yan C, Fu L, Zhang J, Wang J, A comprehensive survey on UAV communication channel modeling 7:07769–107792. https://doi.org/10.1109/ACCESS.2019.2933173. https://ieeexplore.ieee.org/document/8787874/
25. Yildirim I, Basar E, Channel modeling in RIS-empowered wireless communications. https://doi.org/10.48550/ARXIV.2201.06241. arXiv:2201.06241. Publisher: arXiv Version Number: 1

26. You C, Kang Z, Zeng Y, Zhang R, Enabling smart reflection in integrated air-ground wireless network: IRS meets UAV .https://doi.org/10.48550/ARXIV.2103.07151. https://arxiv.org/abs/2103.07151. Publisher: arXiv Version Number: 2
27. Yu X, Xu D, Schober R (2019) MISO wireless communication systems via intelligent reflecting surfaces. In: IEEE/CIC international conference on communications in China (ICCC), pp 735–740. https://doi.org/10.1109/ICCChina.2019.8855810

# Chapter 4
# Physical Layer Security with RIS-Assisted UAV Communication

**Xiaowei Pang, Weidang Lu, and Nan Zhao**

**Abstract** Owing to the high probability of line-of-sight channels, unmanned aerial vehicle (UAV) communications face inherent vulnerabilities to eavesdropping. A promising solution to enhance the security of UAV communication is the integration of reconfigurable intelligent surface (RIS), given its capacity to manipulate propagation environment. This chapter is dedicated to the investigation of secure transmission scheme for RIS-assisted UAV communication networks. Specifically, we introduce an optimization problem aimed at maximizing the average secrecy rate. This problem intricately combines the optimization of transmit beamforming, passive beamforming, and UAV trajectory. To tackle this challenge, we break it down into three distinct sub-problems and propose an iterative algorithm for their alternating resolution. RIS plays a pivotal role in bolstering this security framework, enhancing received signals at intended users while concurrently degrading the quality of eavesdropping links. The efficacy of our proposed scheme is validated through a set of simulations, shedding light on its effectiveness and offering valuable insights. Finally, we highlight several challenges that require resolution for practical implementation in this evolving domain.

**Keywords** Unmanned aerial vehicle · Reconfigurable intelligent surface · Physical layer security · Joint beamforming design · Trajectory optimization

## 4.1 Introduction

Unmanned aerial vehicles (UAVs) have gained significant attraction as aerial communication platforms, due to their high maneuverability and cost-effective deploy-

X. Pang · W. Lu · N. Zhao (✉)
School of Information and Communication Engineering, Dalian University of Technology, Dalian, China
e-mail: zhaonan@dlut.edu.cn
URL: http://www.springer.com/lncs

X. Pang · W. Lu
College of Information Engineering, Zhejiang University of Technology, Hangzhou, China

© The Author(s), under exclusive license to Springer Nature Singapore Pte Ltd. 2026 
I. Ahmad et al. (eds.), *Reconfigurable Intelligent Surfaces for Unmanned Aerial Vehicle Communication*, Unmanned System Technologies,
https://doi.org/10.1007/978-981-96-9662-8_4

ment [1]. The utilization of UAVs is poised to bring enhancements in wireless network coverage, throughput, reliability, and energy efficiency [2–4]. One of the standout advantages of UAVs is their high likelihood of establishing line-of-sight (LoS) wireless links, which greatly improves air-ground communication quality compared to traditional terrestrial communication infrastructure [5]. On the other hand, the air-ground channels face susceptibility to blockages, which can limit the performance of UAV systems [6]. Moreover, the broadcast nature of wireless channels poses security threats for UAV-ground channels, leading to extensive exploration of physical layer security (PLS) techniques in UAV communication systems [7].

Reconfigurable intelligent surface (RIS) have emerged as a promising solution to address these challenges due to their ability to reconfigure wireless propagation environment [8]. Comprising large arrays of reflecting elements, RISs are able to passively reflect signals to enhance desired signals or suppress interference without the need for radio-frequency (RF) chains [9]. Recent research efforts have shifted toward integrating RIS technology into existing UAV networks to unlock their full potential [10, 11]. This integration offers two primary paradigms, depending on the RIS installation. In one scenario, known as the RIS-assisted UAV communication network [12, 13], RIS is placed on buildings to help reflect signals between UAVs and ground nodes. Another intriguing approach is mounting the RIS on UAVs, enabling dynamic reflection from the sky [14]. This configuration leverages UAV mobility and introduces a new degree of freedom for performance improvement.

However, prior works primarily concentrate on minimizing the power or maximizing the sum rate of RIS-assisted UAV systems, with limited attention to addressing eavesdropping threats. In some cases, where legitimate and eavesdropping users share similar channel conditions, conventional beamforming and artificial noise techniques may fall short in ensuring secure transmission. Motivated by these challenges, within this chapter, we propose a secure transmission scheme tailored for RIS-assisted UAV networks. In this scheme, RIS is exploited to enhance legitimate communication quality while degrading eavesdropping quality. Our objective is to maximize the average secrecy rate via jointly optimizing the transmit beamforming and trajectory at the UAV and the passive beamforming at the RIS.

## 4.2 System Model for RIS-Assisted UAV Networks

In this section, we introduce the system configuration and describe the channel models in the RIS-assisted UAV network.

### *4.2.1 System Configuration*

Consider an RIS-assisted UAV network, which is under threat of eavesdropping, as depicted in Fig. 4.1. We assume that the eavesdropper is a network user with-

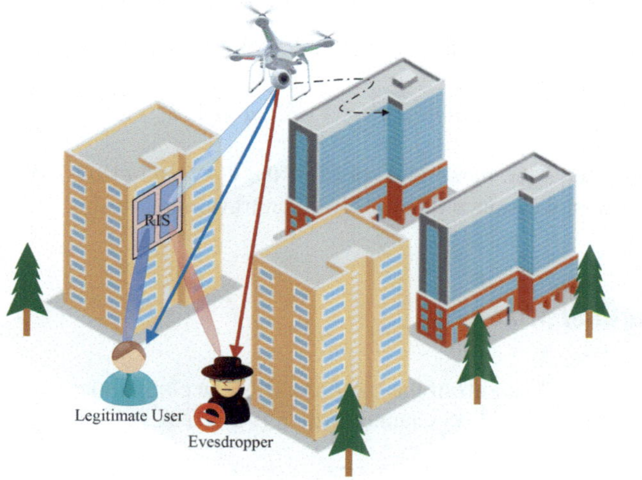


**Fig. 4.1** System model of the RIS-assisted UAV communication network in the presence of eavesdropping

out authorized access to the confidential UAV-to-user communication. Due to the complex urban environment, the direct communication link between the UAV and the ground user is severely obstructed, thereby requiring improvements in channel conditions. To enhance legitimate transmissions while simultaneously countering eavesdropping, an RIS is mounted on the exterior wall of a building to reconfigure the propagation environment. The UAV is equipped with $N_t$ antennas, while both the legitimate user and the eavesdropper have single antennas. The UAV is assumed to fly at a fixed altitude $H_U$ above the ground among a duration $T$. To facilitate subsequent trajectory planning, we discretize the time period $T$ into $N$ time slots, each with a length of $\tau = T/N$. Herein, $\tau$ is chosen to be sufficiently small so that we can approximate the UAV's motion parameters as constant within each time slot. Assume the RIS comprises a uniform planar array (UPA) with $M = M_x M_z$ reflecting elements, where $M_x$ and $M_z$ represent the number of elements along the $x$-axis and $z$-axis, respectively. Denote the phase shift matrix as $\Theta[n] = \text{diag}(e^{j\varpi_1[n]}, \ldots, e^{j\varpi_m[n]}, \ldots, e^{j\varpi_M[n]})$, where $\varpi_m[n]$ represents the phase shift introduced by the $m$th reflecting element in the $n$th time slot.

Under the three-dimensional (3D) Cartesian coordinate system, the horizontal coordinates of the ground legitimate user and the eavesdropper are denoted as $\mathbf{L}_l = [x_l, y_l]^T$ and $\mathbf{L}_e = [x_e, y_e]^T$, respectively. The first element of the RIS serves as a reference point and is assumed to be located at $\mathbf{L}_I = [x_I, y_I]^T$ with a height of $H_I$. The position of the UAV in the $n$th time slot is given by $\mathbf{q}[n] = [x[n], y[n], H_U]^T$ for $n = 1, 2, \ldots, N$. The initial and final locations of the UAV are denoted as $\mathbf{q}_o$ and $\mathbf{q}_f$, respectively. Consequently, it should satisfy the following mobility constraints for the UAV:

$$\mathbf{q}[0] = \mathbf{q}_o, \quad \mathbf{q}[N] = \mathbf{q}_f \ , \tag{4.1}$$

$$\|\mathbf{q}[n] - \mathbf{q}[n-1]\|^2 \leq \hat{V}_M \delta, \quad n = 1, 2, ..., N \ , \tag{4.2}$$

where $\hat{V}_M$ represents the maximum speed of the UAV and $(\hat{V}_M T)/N$ signifies the maximum horizontal distance that the UAV can travel in each slot.

### *4.2.2 Channel Model*

Thanks to the vertical heights of the UAV and RIS, the communication link between them can be modeled as a LoS channel, which is given as

$$\mathbf{H}_{UR}[n] = \sqrt{\beta d_{UR}^{-2}[n]} \mathbf{a}_M^T[n] \mathbf{a}_{N_t}[n] \in \mathbb{C}^{M \times N_t} \ , \tag{4.3}$$

where $d_{UR}[n]$ represents the 3D distance from the UAV to the RIS. $\mathbf{a}_M^T[n]$ and $\mathbf{a}_{N_t}[n]$ denote the receive array response of RIS and the transmit array response of UAV during the $n$th time slot, respectively.

In this complex urban environment, obstacles such as buildings and trees block the LoS link from the UAV to the ground nodes. Consequently, direct channels consist of numerous scattering components, leading to Rayleigh fading in the UAV-ground channels. The channel model from the UAV to the legitimate user and the eavesdropper is given as

$$\mathbf{h}_{u,k}^H[n] = \sqrt{\beta d_{u,k}^{-\alpha}[n]} \bar{\mathbf{h}}_{u,k}[n] \in \mathbb{C}^{1 \times N_t}, k \in \{l, e\} \ , \tag{4.4}$$

where $\beta$ represents the channel power gain at the reference distance $d_0 = 1$ m, $\alpha$ is the path-loss exponent, and $d_{u,i}[n]$ is the distance from the UAV to the secure user or eavesdropper in the $n$th time slot, with $k \in l, e$. The elements of $\bar{\mathbf{h}}_{u,k}[n] (k \in l, e)$ are independent and identically distributed complex Gaussian variables with zero mean and unit variance.

Regarding the RIS-ground channels, they include a combination of LoS and Non-Line-of-Sight (NLoS) elements, giving rise to a Rician fading model. The channel from the RIS to a ground node $k$, $k \in l, e$, is modeled as

$$\mathbf{h}_{R,k}[n] = \sqrt{\beta d_{R,k}^{-\tilde{\alpha}}} \left( \sqrt{\frac{\mathcal{K}}{\mathcal{K}+1}} \bar{\mathbf{h}}_{R,k}^L[n] + \sqrt{\frac{1}{\mathcal{K}+1}} \bar{\mathbf{h}}_{R,k}^N[n] \right) \ , \tag{4.5}$$

where $d_{R,k}$ represents the distance between the RIS and the ground node, $\tilde{\alpha}$ is the path-loss exponent, and $\mathcal{K}$ denotes the Rician factor. The LoS component is obtained with azimuth and elevation AoDs for the LoS link from the RIS to node $k$ denoted by

$\phi_k$ and $\theta_k$. The NLoS component $\bar{\mathbf{h}}^N_{R,k}[n] \in \mathbb{C}^{M\times 1}$ is modeled as complex Gaussian distributed with zero mean and unit variance, i.e., $\bar{\mathbf{h}}^N_{R,k}[n] \sim \mathcal{CN}(\mathbf{0}, \mathbf{I})$.

As a result, the signal received by user $k$ ($k \in l, e$) in the $n$th time slot can be written as

$$y_i[n] = \left(\mathbf{h}^H_{u,k}[n] + \mathbf{h}_{R,k}[n]\Theta[n]\mathbf{H}_{UR}[n]\right)\mathbf{v}[n]x[n] + z_k[n] \,, \tag{4.6}$$

where $x[n]$ represents the transmitted signal, $\mathbf{v}[n] \in \mathbb{C}^{N_t\times 1}$ denotes the transmit beamforming vector, and $z_k[n] \sim \mathcal{CN}\left(0, \sigma_k^2\right)$ stands for the additive white Gaussian noise (AWGN).

*Remark*: In this chapter, we make the assumption that all channel state information (CSI) are available, which is reasonable for two considerations. First, in scenarios where the eavesdropper also acts as a licensed user but lacks the authorization to access confidential information, the transmitter can leverage perfect CSI obtained from the eavesdropper. Second, designing optimization algorithms with perfect CSI sets a performance upper bound for cases with imperfect CSI, such as statistical CSI. Benefiting from well-established RIS-assisted network channel estimation algorithms [15], our research focuses on addressing the optimization problem effectively.

## 4.3 Joint Design for RIS-Assisted Secure UAV Communication

In this section, we propose a joint transmission design for RIS-assisted secure UAV communications by solving the formulated secrecy rate maximization problem. Furthermore, an iterative algorithm is developed to optimize the three variables alternately.

### *4.3.1 Problem Formulation*

First, the achievable rate of the legitimate user and that of the eavesdropper in the $n$th time slot can be, respectively, expressed as

$$R_l[n] = \log_2\left(1 + \frac{\left|\left(\mathbf{h}^H_{u,l}[n] + \mathbf{h}^H_{R,l}[n]\Theta[n]\mathbf{H}_{UR}[n]\right)\mathbf{v}[n]\right|^2}{\sigma_l^2}\right), \tag{4.7}$$

$$R_e[n] = \log_2\left(1 + \frac{\left|\left(\mathbf{h}^H_{u,e}[n] + \mathbf{h}^H_{R,e}[n]\Theta[n]\mathbf{H}_{UR}[n]\right)\mathbf{v}[n]\right|^2}{\sigma_e^2}\right). \tag{4.8}$$

Then, the average achievable secrecy rate at the legitimate user is given as

$$\bar{R}_S = \frac{1}{N}\sum_{n=1}^{N}\left[R_l[n] - R_e[n]\right]^+ , \tag{4.9}$$

where $[a]^+ \triangleq \max(a, 0)$. Note that we can always ensure a nonnegative secrecy rate by adjusting the transmit power. Consequently, the operator $[\cdot]^+$ can be safely removed without altering the value of (4.9).

To ensure secure transmission, our objective is to maximize the average achievable secrecy rate by jointly optimizing three variables: the transmit beamforming $\mathbf{V} \triangleq \{\mathbf{v}[n], \forall n\}$, the phase shift matrix $\Psi \triangleq \{\Theta[n], \forall n\}$, and the UAV trajectory $\mathbf{Q} \triangleq \{\mathbf{q}[n], \forall n\}$. The formulated problem is presented as

$$\begin{aligned} \max_{\mathbf{V},\Psi,\mathbf{Q}} \quad & \bar{R}_S \\ s.t. \quad & \mathbf{q}[0] = \mathbf{q}_o, \quad \mathbf{q}[N] = \mathbf{q}_f , \\ & \|\mathbf{q}[n] - \mathbf{q}[n-1]\|^2 \le \hat{V}_M\delta, \quad n = 1, 2, ..., N , \\ & 0 \le \varpi_m[n] \le 2\pi, \ \forall m, \forall n , \\ & \|\mathbf{v}[n]\|^2 \le P_U, \ \forall n , \end{aligned} \tag{4.10}$$

where $P_U$ represents the maximum transmit power of the UAV. Solving this optimization problem is challenging due to the couple of variables and non-convex constraints. To tackle this problem, we break down it into three sub-problems and solve them alternately.

### 4.3.2 Transmit Beamforming Optimization

Fixing $\Psi$ and $\mathbf{Q}$, we can rewrite the transmit beamforming optimization problem as

$$\max_{\mathbf{v}[n]} \frac{|\bar{\mathbf{h}}_l[n]\mathbf{v}[n]|^2 + \sigma_l^2}{|\bar{\mathbf{h}}_e[n]\mathbf{v}[n]|^2 + \sigma_e^2} \tag{4.11}$$

$$s.t. \quad \|\mathbf{v}[n]\|^2 \le P_U, \ \forall n , \tag{4.12}$$

where $\bar{\mathbf{h}}_k[n]$ denotes the concatenated channel and is represented as $\bar{\mathbf{h}}_k[n] = \mathbf{h}_{u,k}^H[n] + \mathbf{h}_{R,k}^H[n]\Theta[n]\mathbf{H}_{UR}[n]$, $k \in \{l, e\}$. Define $\bar{\mathbf{H}}_k[n] = \bar{\mathbf{h}}_k^H[n]\bar{\mathbf{h}}_k[n]$, (4.11) can be further written as

$$\max_{\mathbf{v}[n]} \frac{\mathbf{v}^H[n]\bar{\mathbf{H}}_l[n]\mathbf{v}[n] + \sigma_l^2}{\mathbf{v}^H[n]\bar{\mathbf{H}}_e[n]\mathbf{v}[n] + \sigma_e^2} . \tag{4.13}$$

Then, we can directly determine the optimal solution to transmit beamforming as $\mathbf{v}_{opt}[n] = \sqrt{P_U}\mathbf{v}_{max}[n]$, where $\mathbf{v}_{max}[n]$ corresponds to the eigenvector associated with the largest eigenvalue of the matrix $\left(\bar{\mathbf{H}}_e[n]P_U + \sigma_e^2\mathbf{I}_{N_t}\right)^{-1}\left(\bar{\mathbf{H}}_l[n]P_U + \sigma_l^2\mathbf{I}_{N_t}\right)$.

### *4.3.3 Passive Beamforming Optimization*

Rewrite $\Theta[n]$ as $\Theta[n] = \text{diag}(\theta_1[n], ..., \theta_M[n])$ with unit modulus constraints. With optimized transmit beamforming and given UAV trajectory, we aim to find the passive beamforming of the RIS by solving the following problem:

$$\begin{aligned}\max_{\Theta[n]} \quad & \frac{\left|\left(\mathbf{h}_{u,l}^H[n] + \mathbf{h}_{R,l}^H[n]\Theta[n]\mathbf{H}_{UR}[n]\right)\mathbf{v}[n]\right|^2 + \sigma_l^2}{\left|\left(\mathbf{h}_{u,e}^H[n] + \mathbf{h}_{R,e}^H[n]\Theta[n]\mathbf{H}_{UR}[n]\right)\mathbf{v}[n]\right|^2 + \sigma_e^2} \\ s.t. \quad & |\theta_m[n]| = 1, \ m = 1, ..., M .\end{aligned} \tag{4.14}$$

Then, by introducing an auxiliary vector $\mathbf{g}^H[n] = [\theta_1[n], ..., \theta_M[n]]$, we have $\mathbf{h}_{R,l}^H[n]\Theta[n]\mathbf{H}_{UR}[n]\mathbf{v}[n] = \mathbf{g}^H[n]\text{diag}\{\mathbf{h}_{R,l}^H[n]\}\mathbf{H}_{UR}[n]\mathbf{v}[n]$. Then, we can rewrite the objective function as

$$\max_{\mathbf{g}[n]} \frac{\left|\mathbf{a}_l[n] + \mathbf{g}^H[n]\mathbf{b}_l[n]\right|^2 + \sigma_l^2}{\left|\mathbf{a}_e[n] + \mathbf{g}^H[n]\mathbf{b}_e[n]\right|^2 + \sigma_e^2}, \tag{4.15}$$

where $\mathbf{a}_l[n] = \mathbf{h}_{u,l}^H[n]\mathbf{v}[n]$, $\mathbf{a}_e[n] = \mathbf{h}_{u,e}^H[n]\mathbf{v}[n]$, $\mathbf{b}_l[n] = \text{diag}\{\mathbf{h}_{R,l}^H[n]\}\mathbf{H}_{UR}[n]\mathbf{v}[n]$, and $\mathbf{b}_e[n] = \text{diag}\{\mathbf{h}_{R,e}^H[n]\}\mathbf{H}_{UR}[n]\mathbf{v}[n]$. Equation (4.15) is further transformed into an equivalent minimization problem as

$$\min_{\mathbf{g}[n]} \frac{\left|\mathbf{a}_e[n] + \mathbf{g}^H[n]\mathbf{b}_e[n]\right|^2 + \sigma_e^2}{\left|\mathbf{a}_l[n] + \mathbf{g}^H[n]\mathbf{b}_l[n]\right|^2 + \sigma_l^2}. \tag{4.16}$$

Thus, it can be reformulated as a sequence of parametric sub-problems:

$$\begin{aligned}\min_{\mathbf{g}[n]} \quad & \left|\mathbf{a}_e[n] + \mathbf{g}^H[n]\mathbf{b}_e[n]\right|^2 + \sigma_e^2 - \xi^{(r-1)}\left(\left|\mathbf{a}_l[n] + \mathbf{g}^H[n]\mathbf{b}_l[n]\right|^2 + \sigma_l^2\right) \\ s.t. \quad & |\theta_m[n]| = 1, \ m = 1, ..., M ,\end{aligned} \tag{4.17}$$

where $\xi^{(r)}(r \geq 0)$ is a nonnegative parameter. With an initial value $\xi^{(0)} = 0$, $\xi^{(r)}$ can be updated as

$$\xi^{(r)} = \frac{\left|\mathbf{a}_e[n] + \mathbf{g}^{(r)H}[n]\mathbf{b}_e[n]\right|^2 + \sigma_e^2}{\left|\mathbf{a}_l[n] + \mathbf{g}^{(r)H}[n]\mathbf{b}_l[n]\right|^2 + \sigma_l^2}, \tag{4.18}$$

where $\mathbf{g}^{(r)H}[n]$ represents the solution obtained in the $r$th iteration. Due to the fact that (4.18) remains non-convex, we replace it with a tractable upper bound as

$$\begin{aligned}&\left|\mathbf{a}_e[n]+\mathbf{g}^H[n]\mathbf{b}_e[n]\right|^2+\sigma_e^2-\xi\left(\left|\mathbf{a}_l[n]+\mathbf{g}^H[n]\mathbf{b}_l[n]\right|^2+\sigma_l^2\right)\\&\leq\lambda_{\max}(\tilde{\mathbf{A}}[n])\left\|\mathbf{g}[n]\right\|^2-2\mathrm{Re}\left\{\mathbf{g}^H[n]\tilde{\mathbf{B}}[n]\right\}+\tilde{\mathbf{C}}[n]\,,\end{aligned} \tag{4.19}$$

$\tilde{\mathbf{A}}[n]$, $\tilde{\mathbf{B}}[n]$, and $\tilde{\mathbf{C}}[n]$ are expressed as

$$\tilde{\mathbf{A}}[n]=\mathbf{b}_e[n]\mathbf{b}_e^H[n]-\xi\mathbf{b}_l[n]\mathbf{b}_l^H[n]\,, \tag{4.20}$$

$$\tilde{\mathbf{B}}[n]=\left(\lambda_{\max}(\tilde{\mathbf{A}}[n])\mathbf{I}-\tilde{\mathbf{A}}[n]\right)\tilde{\mathbf{g}}[n]+\xi\mathbf{a}_l^*[n]\mathbf{b}_l[n]-\mathbf{a}_e^*[n]\mathbf{b}_e[n]\,, \tag{4.21}$$

$$\tilde{\mathbf{C}}[n]=\tilde{\mathbf{g}}^H[n]\left(\lambda_{\max}(\tilde{\mathbf{A}}[n])\mathbf{I}-\tilde{\mathbf{A}}[n]\right)\tilde{\mathbf{g}}[n]+\left|\mathbf{a}_e[n]\right|^2+\sigma_e^2-\xi\left|\mathbf{a}_l[n]\right|^2-\xi\sigma_l^2\,. \tag{4.22}$$

where $\tilde{\mathbf{g}}[n]$ denotes the optimal solution to $\mathbf{g}[n]$ in the previous iteration. As a result, the objective function can be rewritten as

$$\min_{\mathbf{g}[n]}\ \lambda_{\max}(\tilde{\mathbf{A}}[n])\left\|\mathbf{g}[n]\right\|^2-2\mathrm{Re}\left\{\mathbf{g}^H[n]\tilde{\mathbf{B}}[n]\right\}\,. \tag{4.23}$$

Due to the unit modulus constraint $|\theta_m[n]|=1$, we have $\|\mathbf{g}[n]\|^2=M$. In this context, the objective is equivalent to maximizing $\mathrm{Re}\left\{\mathbf{g}^H[n]\tilde{\mathbf{B}}[n]\right\}$. As a result, the optimal passive beamforming vector of RIS can be determined as

$$\mathbf{g}_{opt}^H[n]=\left[e^{j\arg(\tilde{\mathbf{B}}_1[n])},\ ...,\ e^{j\arg(\tilde{\mathbf{B}}_M[n])}\right]\,, \tag{4.24}$$

where $\tilde{\mathbf{B}}_m[n]$ denotes $m$th element entry of $\tilde{\mathbf{B}}[n]$.

### 4.3.4 UAV Trajectory Optimization

Fixing the transmit and passive beamforming vectors as obtained above, we can express the UAV trajectory optimization problem as

$$\begin{aligned}\max_{\mathbf{Q}}\ &\frac{1}{N}\sum_{n=1}^{N}\left(R_l[n]-R_e[n]\right)\\s.t.\ &\mathbf{q}[0]=\mathbf{q}_o,\ \ \mathbf{q}[N]=\mathbf{q}_f\,,\\&\left\|\mathbf{q}[n]-\mathbf{q}[n-1]\right\|^2\leq\hat{V}_M\delta,\ \ n=1,2,...,N\,.\end{aligned} \tag{4.25}$$

It is worth noting that both $R_l[n]$ and $R_e[n]$ are non-convex and non-concave. To address $R_l[n]$, we recast the received signal power at the legitimate user as

$$\begin{aligned}&\left|\left(\mathbf{h}_{u,l}^H[n]+\mathbf{h}_{R,l}^H[n]\Theta[n]\mathbf{H}_{UR}[n]\right)\mathbf{v}[n]\right|^2\\&=\frac{\beta}{\sigma^2}\left(X_l[n]d_{u,l}^{-\alpha}[n]+Y_l[n]d_{u,l}^{-\frac{\alpha}{2}}[n]d_{UR}^{-1}[n]+Z_l[n]d_{UR}^{-2}[n]\right).\end{aligned} \tag{4.26}$$

The coefficients $X_l[n]$, $Y_l[n]$, and $Z_l[n]$ are determined by

$$X_l[n]=\bar{\mathbf{h}}_{u,l}[n]\mathbf{V}[n]\bar{\mathbf{h}}_{u,l}^H[n], \tag{4.27}$$

$$Y_l[n]=2\mathrm{Re}\left\{\mathbf{G}_l[n]\mathbf{a}_M^T[n]\mathbf{a}_{N_t}[n]\mathbf{V}[n]\bar{\mathbf{h}}_{u,l}^H[n]\right\}, \tag{4.28}$$

$$Z_l[n]=\mathbf{G}_l[n]\mathbf{a}_M^T[n]\mathbf{a}_{N_t}[n]\mathbf{V}[n]\mathbf{a}_{N_t}^*[n]\mathbf{a}_{N_t}^H[n]\mathbf{G}_l^H[n], \tag{4.29}$$

where $\mathbf{V}[n]=\mathbf{v}[n]\mathbf{v}^H[n]$ and $\mathbf{G}_l[n]=\mathbf{h}_{R,l}^H[n]\Theta[n]$.

The expression for $R_e[n]$ can be transformed into a similar form as that of $R_l[n]$. To simplify (4.25), we introduce auxiliary variables $\mathbf{t}_1=\{t_1[n],\forall n\}$ and $\mathbf{t}_2=\{t_2[n],\forall n\}$, and reformulate it as

$$\begin{aligned}\max_{\substack{\mathbf{Q},\mathbf{t}_1,\mathbf{t}_2,\\ \mathbf{u}_1,\mathbf{u}_2,\mathbf{u}_3}}\quad&\frac{1}{N}\sum_{n=1}^{N}\left(\log_2(e)t_1[n]-\log_2(e)t_2[n]\right)\\ s.t.\quad&1+F_L[n]\beta/\sigma^2\geq e^{t_1[n]},\\&1+F_E[n]\beta/\sigma^2\leq e^{t_2[n]},\\&\mathbf{q}[0]=\mathbf{q}_o,\quad\mathbf{q}[N]=\mathbf{q}_f,\\&\|\mathbf{q}[n]-\mathbf{q}[n-1]\|^2\leq\hat{V}_M\delta,\quad n=1,2,\ldots,N.\end{aligned} \tag{4.30}$$

Herein, $F_L[n]$ and $F_E[n]$ are expressed as

$$F_L[n]=X_l[n]d_{u,l}^{-\alpha}[n]+Y_l[n]d_{u,l}^{-\frac{\alpha}{2}}[n]d_{UR}^{-1}[n]+Z_l[n]d_{UR}^{-2}[n], \tag{4.31}$$

$$F_E[n]=X_e[n]d_{u,e}^{-\alpha}[n]+Y_e[n]d_{u,e}^{-\frac{\alpha}{2}}[n]d_{UR}^{-1}[n]+Z_e[n]d_{UR}^{-2}[n]. \tag{4.32}$$

Particularly, $F_L[n]$ and $F_E[n]$ should be converted into a concave and a convex ones, respectively.

To handle $F_L[n]$, we introduce several new functions as

$$\mathcal{S}_x[n]=X_l[n]d_{u,l}^{-\alpha}[n], \tag{4.33}$$

$$\mathcal{S}_y[n]=Y_l[n]d_{u,l}^{-\frac{\alpha}{2}}[n]d_{UR}^{-1}[n], \tag{4.34}$$

$$\mathcal{S}_z[n]=Z_l[n]d_{UR}^{-2}[n]. \tag{4.35}$$

This modification simplifies the optimization problem, and to ensure convexity, we implement the following transformations.

Initially, we introduce slack variables $\mathbf{u}_1 = \{u_1[n], \forall n\}$ and $\mathbf{u}_3 = \{u_3[n], \forall n\}$, along with the following constraints:

$$d_{u,l}[n] \leq u_1[n], \quad d_{UR}[n] \leq u_3[n], \ \forall n \ . \tag{4.36}$$

Given that $u_1^2[n]$ is convex, with the given local point $\bar{u}_1[n]$, it yields

$$d_{u,l}^2[n] + \bar{u}_1^2[n] - 2\bar{u}_1[n]u_1[n] \leq 0, \ \forall n \ . \tag{4.37}$$

Considering that $\mathbf{V}[n]$ is typically a positive semi-definite matrix, it follows that $X_l[n] \geq 0$ and $Z_l[n] \geq 0$. By replacing $d_{u,l}[n]$ and $d_{UR}[n]$ with $u_1[n]$ and $u_3[n]$, respectively, we have

$$\mathcal{S}_x[n] \geq X_l[n]u_1^{-\alpha}[n] \ , \qquad \mathcal{S}_z[n] \geq Z_l[n]u_3^{-2}[n] \ , \tag{4.38}$$

both of which are convex with respect to $u_1[n]$ and $u_3[n]$. Thus, $\mathcal{S}_x[n]$ and $\mathcal{S}_z[n]$ have lower bounds given by their respective first-order Taylor series of the right-hand terms as

$$\mathcal{S}_x[n] \geq (1+\alpha)X_l[n]\bar{u}_1^{-\alpha}[n] - \alpha X_l[n]\bar{u}_1^{-\alpha-1}[n]u_1[n] \triangleq \Upsilon_x\left(\bar{u}_1[n], u_1[n]\right) \ , \tag{4.39}$$

$$\mathcal{S}_z[n] \geq 3Z_l[n]\bar{u}_3^{-2}[n] - 2Z_l[n]\bar{u}_3^{-3}[n]v_3[n] \triangleq \Upsilon_z\left(\bar{u}_3[n], u_3[n]\right) \ . \tag{4.40}$$

However, $Y_l[n]$ may not necessarily be positive or negative. When $Y_l[n] \geq 0$, we can approximate $\mathcal{S}_y[n]$ using slack variables $u_1[n]$ and $u_3[n]$ as

$$\begin{aligned} \mathcal{S}_y[n] \geq\ & (2+\frac{\alpha}{2})Y_l[n]\bar{u}_1^{-\frac{\alpha}{2}}[n]\bar{u}_3^{-1}[n] \\ & - \frac{\alpha}{2}Y_l[n]\bar{u}_3^{-1}[n]\bar{u}_1^{-\frac{\alpha}{2}-1}[n]v_1[n] - Y_l[n]\bar{u}_1^{-\frac{\alpha}{2}}[n]\bar{u}_3^{-2}[n]u_3[n] \\ & \triangleq \Upsilon_y\left(\bar{u}_1[n], \bar{u}_3[n], u_1[n], u_3[n]\right) \ . \end{aligned} \tag{4.41}$$

Conversely, if $Y_l[n] < 0$, we introduce two additional relaxation variables $\mathbf{U}_1 = \{U_1[n], \forall n\}$ and $\mathbf{U}_3 = \{U_3[n], \forall n\}$, satisfying

$$U_1[n] \leq d_{u,l}[n], \quad U_3[n] \leq d_{UR}[n], \ \forall n \ . \tag{4.42}$$

In this context, $Y_l[n]U_1^{-\frac{\alpha}{2}}[n]U_3^{-1}[n]$ serves as a lower bound for $\mathcal{S}_y[n]$. Subsequently, with the given local points $\bar{U}_1[n]$ and $\bar{U}_3[n]$, the lower bound of $\mathcal{S}_y[n]$ can be expressed as

$$\mathcal{S}_y[n] \geq \frac{Y_l[n]}{2}\left[\left(U_1^{-\frac{\alpha}{2}}[n] + U_3^{-1}[n]\right)^2 - U_1^{-\alpha} - U_3^{-2}[n]\right]$$
$$\geq \frac{Y_l[n]}{2}\Big[\left(U_1^{-\frac{\alpha}{2}}[n] + U_3^{-1}[n]\right)^2 - (1+\alpha)\bar{U}_1^{-\alpha}[n]$$
$$+ \alpha\bar{U}_1^{-\alpha-1}[n]U_1[n] - 3\bar{U}_3^{-2}[n] + 2\bar{U}_3^{-3}[n]U_3[n]\Big]$$
$$\triangleq \Gamma_y\left(\bar{U}_1[n], \bar{U}_3[n], U_1[n], U_3[n]\right) , \tag{4.43}$$

where $\Gamma_y$ can be readily shown to be concave under the condition $\mathcal{S}_y[n] < 0$ by analyzing its Hessian matrix.

Since the negative norm-squared function $-\|\mathbf{q}[n] - \mathbf{L}_l\|^2$ is concave with respect to the variable vector $\mathbf{q}[n]$, it possesses a convex upper bound as

$$-\|\mathbf{q}[n] - \mathbf{L}_l\|^2 \leq \left\|\mathbf{q}^r[n]\right\|^2 - 2(\mathbf{q}^r[n] - \mathbf{L}_l)^T\mathbf{q}[n] - \|\mathbf{L}_l\|^2 , \tag{4.44}$$

where $\mathbf{q}^r[n]$ denotes the given point and $\mathbf{Q}^r \triangleq \{\mathbf{q}^r[n], \forall n\}$. Hence, by replacing $-d_{u,l}^2[n]$ with its upper bound, we have

$$U_1^2[n] + \left\|\mathbf{q}^r[n]\right\|^2 - 2(\mathbf{q}^r[n] - \mathbf{L}_l)^T\mathbf{q}[n] - \|\mathbf{L}_l\|^2 + H_U^2 \leq 0 . \tag{4.45}$$

Furthermore, we define a binary variable $I_y[n]$ to indicate that $Y_l[n] \geq 0$ in the $n$th time slot if $I_y[n] = 1$; otherwise, $I_y[n] = 0$. We can derive a concave lower bound for $F_L[n]$ as

$$F_L[n] = \mathcal{S}_x[n] + \mathcal{S}_y[n] + \mathcal{S}_z[n]$$
$$\geq \Upsilon_x[n] + \Upsilon_z[n] + I_y[n]\Upsilon_y[n] + (1 - I_y[n])\Gamma_y[n]$$
$$\triangleq \widetilde{F}_D[n] . \tag{4.46}$$

In (4.30), $e^{t_2[n]}$ is convex, and therefore we can approximate it using its first-order Taylor expansion at a given feasible point $\bar{t}_2[n]$ as

$$e^{t_2[n]} \geq e^{\bar{t}_2[n]}(t_2[n] - \bar{t}_2[n] + 1) . \tag{4.47}$$

We also introduce new functions related to $F_E[n]$ as follows:

$$\tilde{\mathcal{S}}_x[n] = X_e[n]d_{u,e}^{-\alpha}[n] , \tag{4.48}$$
$$\tilde{\mathcal{S}}_z[n] = Z_e[n]d_{UR}^{-2}[n] , \tag{4.49}$$
$$\tilde{\mathcal{S}}_y[n] = Y_e[n]d_{u,e}^{-\frac{\alpha}{2}}[n]d_{UR}^{-1}[n] . \tag{4.50}$$

Unlike $F_L[n]$ that requires a lower bound, $F_E[n]$ needs to be approximated by a convex upper bound. To this end, auxiliary variables $\mathbf{u}_2 = \{u_2[n], \forall n\}$ and $\mathbf{U}_2 = \{U_2[n], \forall n\}$ are introduced as

$$d_{u,e}[n] \le u_2[n], \quad U_2[n] \le d_{u,e}[n], \ \forall n \ . \tag{4.51}$$

Then, we substitute $d_{u,e}[n]$ and $d_{UR}[n]$ with $U_2[n]$ and $U_3[n]$, respectively, to obtain their convex approximations as

$$\tilde{\mathcal{S}}_x[n] \le X_e[n] V_2^{-\alpha}[n] \triangleq \mathcal{F}_x[n] \ , \tag{4.52}$$

$$\tilde{\mathcal{S}}_z[n] \le Z_e[n] V_3^{-2}[n] \triangleq \mathcal{F}_z[n] \ . \tag{4.53}$$

For the case of $Y_e[n] \ge 0$, $\tilde{\mathcal{S}}_y[n]$ is approximated by its convex counterpart as

$$\begin{aligned} \tilde{\mathcal{S}}_y[n] \le & \frac{Y_e[n]}{2}\Big[ \left(U_2^{-\frac{\alpha}{2}}[n] + U_3^{-1}[n]\right)^2 - (1+\alpha)\bar{U}_2^{-\alpha}[n] \\ & + \alpha \bar{U}_2^{-\alpha-1}[n] U_2[n] - 3\bar{U}_3^{-2}[n] + 2\bar{U}_3^{-3}[n] U_3[n] \Big] \\ & \triangleq \mathcal{F}_y\left(\bar{U}_2[n], \bar{U}_3[n], U_2[n], U_3[n]\right) \ , \end{aligned} \tag{4.54}$$

where $\bar{U}_2[n]$ and $\bar{U}_3[n]$ are the given local points.

When $Y_e[n] < 0$, we proceed to approximate $\tilde{\mathcal{S}}_y[n]$ using its upper-bound function with respect to $u_2[n]$ and $u_3[n]$. Specifically, we convert it into a convex form as

$$\begin{aligned} \tilde{\mathcal{S}}_y[n] \le & (2 + \frac{\alpha}{2}) Y_e[n] \bar{u}_2^{-\frac{\alpha}{2}}[n] \bar{u}_3^{-1}[n] \\ & - \frac{\alpha}{2} Y_e[n] \bar{u}_3^{-1}[n] \bar{u}_2^{-\frac{\alpha}{2}-1}[n] v_2[n] - Y_e[n] \bar{u}_2^{-\frac{\alpha}{2}}[n] \bar{u}_3^{-2}[n] u_3[n] \\ & \triangleq \mathcal{Q}_y\left(\bar{u}_2[n], \bar{u}_3[n], u_2[n], u_3[n]\right) \ , \end{aligned} \tag{4.55}$$

where $\bar{u}_2[n]$ and $\bar{u}_3[n]$ represent the given feasible points.

Using $\mathcal{I}_y[n] = 1$ to imply $Y_e[n] \ge 0$, $F_E[n]$ can be approximated as

$$\begin{aligned} F_E[n] = & \tilde{\mathcal{S}}_x[n] + \tilde{\mathcal{S}}_y[n] + \tilde{\mathcal{S}}_z[n] \\ & \le \mathcal{F}_x[n] + \mathcal{F}_z[n] + \mathcal{I}_y[n] \mathcal{F}_y[n] + (1 - \mathcal{I}_y[n]) \mathcal{Q}_y[n] \\ & \triangleq \widetilde{F}_E[n] \ . \end{aligned} \tag{4.56}$$

By introducing these approximations and transformations, we are able to reformulate the trajectory optimization problem into a convex problem as

$$
\begin{aligned}
\max_{\substack{\mathbf{Q},\mathbf{t}_1,\mathbf{t}_2,\\ \mathbf{u}_1,\mathbf{u}_2,\mathbf{u}_3\\ \mathbf{U}_1,\mathbf{U}_2,\mathbf{U}_3}} \quad & \frac{1}{N}\sum_{n=1}^{N} t_1[n]-t_2[n] \\
s.t. \quad & 1+\frac{\beta}{\sigma^2}\widetilde{F}_D[n] \ge e^{t_1[n]} , \\
& 1+\frac{\beta}{\sigma^2}\widetilde{F}_E[n] \le e^{\bar{t}_2[n]}(t_2[n]-\bar{t}_2[n]+1) , \\
& d_{u,l}^2[n]+\bar{u}_1^2[n]-2\bar{u}_1[n]u_1[n] \le 0,\ \forall n , \\
& d_{u,e}^2[n]+\bar{u}_2^2[n]-2\bar{u}_2[n]u_2[n] \le 0,\ \forall n , \\
& d_{UR}^2[n]+\bar{u}_3^2[n]-2\bar{u}_3[n]u_3[n] \le 0,\ \forall n , \\
& U_1^2[n]+\left\|\mathbf{q}^r[n]\right\|^2-2(\mathbf{q}^r[n]-\mathbf{L}_l)^T\mathbf{q}[n]-\|\mathbf{L}_l\|^2+H_U^2 \le 0 , \\
& U_2^2[n]+\left\|\mathbf{q}^r[n]\right\|^2-2(\mathbf{q}^r[n]-\mathbf{L}_e)^T\mathbf{q}[n]-\|\mathbf{L}_e\|^2+H_U^2 \le 0 , \\
& U_3^2[n]+\left\|\mathbf{q}^r[n]\right\|^2-2(\mathbf{q}^r[n]-\mathbf{L}_I)^T\mathbf{q}[n]-\|\mathbf{L}_I\|^2+H_{uI}^2 \le 0 , \\
& \mathbf{q}[0]=\mathbf{q}_o,\quad \mathbf{q}[N]=\mathbf{q}_f , \\
& \|\mathbf{q}[n]-\mathbf{q}[n-1]\|^2 \le \hat{V}_M\delta,\quad n=1,2,...,N .
\end{aligned} \tag{4.57}
$$

This convex problem can be efficiently solved using existing optimization tools such as CVX.

### 4.3.5 Overall Algorithm

To solve the original joint optimization problem (4.10), we introduce an effective algorithm tackling the three sub-problems iteratively until reaching convergence. Without loss of generality, the algorithm is based on the concept of block coordinate descent, where the three sub-problems are solved in blocks while keeping the other two variables fixed. Specifically, with fixed UAV trajectory and RIS phase shifts, we could directly derive the optimal transmit beamforming. By contrast, the solutions to RIS passive beamforming and UAV trajectory are obtained by addressing approximate problems (4.23) and (4.57), stemming from their non-convex counterparts in each iteration. It should be noted that $\bar{R}_S$, computed by solving (4.23) or (4.57), generally establishes a lower bound for (4.10). Since the objective value of the original problem (4.10) has an upper bound, and the solutions to the various sub-problems can gradually approach the optimal solution of the original problem through iteration, the overall algorithm ultimately converges to at least a locally optimal solution. The specific details of the algorithm are not provided in this chapter, and readers are encouraged to refer to [16] for more information.

## 4.4 Numerical Results and Discussions

In this section, we present simulations to evaluate the performance of our proposed joint design for RIS-assisted UAV networks. Default simulation parameters are given as follows: $H_U = 100$ m, $\hat{V}_M = 20$ m/s, $H_I = 30$ m, $N_t = 16$, $M = 64$, $P_U = 20$ dBm, $\beta_0 = -30$ dB, and $\sigma_l^2 = \sigma_e^2 = -80$ dBm. In addition, we set $\alpha = 3.5$ and $\tilde{\alpha} = 2.4$. The coordinate system is centered at the RIS location. The legitimate user is positioned at $\mathbf{L}_l = [20, 30]^T$ m, while the UAV's initial and final horizontal coordinates are $\mathbf{q}_o = [-300, 80]^T$ m and $\mathbf{q}_f = [300, 80]^T$ m, respectively. In order to achieve effective signal reflection, it is assumed that both the UAV and the legitimate user are located on the same side of the RIS. Considering a less favorable scenario, we assume that the eavesdropper's position is randomly generated in proximity to the legitimate user.

### 4.4.1 *Optimized UAV Trajectory*

Figure 4.2 displays the spatial arrangement of legitimate and eavesdropping users as well as RIS, along with the optimized flight trajectories of the UAV in various scenarios. We refer to the baseline scenario as "Without RIS," where both transmit beamforming and UAV trajectory are jointly optimized, and no RIS is employed. Given the specified initial and final locations, it is evident that the UAV must dedicate

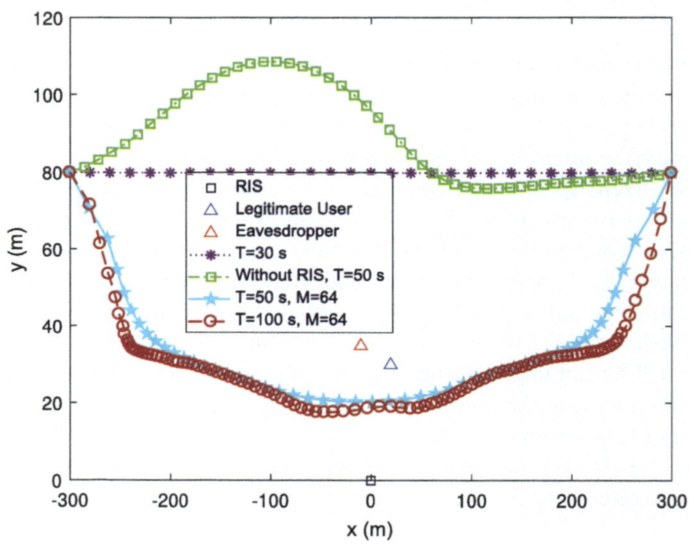


**Fig. 4.2** Distribution of network nodes and UAV trajectories under different scenarios

a minimum of $T = 30$ s to travel directly from its initial position to the destination at its maximum speed. In the absence of RIS, the UAV's flight pattern tends to divert away from the eavesdropper, aimed at averting potential information leakage. This deviation in trajectory is caused by the eavesdropper's close proximity to the legitimate user.

### 4.4.2 *Secrecy Performance Comparison*

In Fig. 4.3, we compare the proposed scheme with two other benchmarks in terms of average secrecy rate for different values of $T$. It can be observed that the secrecy rate of the schemes with trajectory optimization increases with $T$ and eventually stabilizes. This is because the UAV has more time to serve legitimate users in favorable locations with larger $T$. By contrast, the secrecy rate remains nearly constant without trajectory optimization due to the uniform mobility of the UAV and the average operation on the rate. It is worth noting that the gaps between the proposed scheme and other schemes demonstrate the performance improvement caused by the joint optimization of the transmit beamforming, the phase shift, and the UAV trajectory.

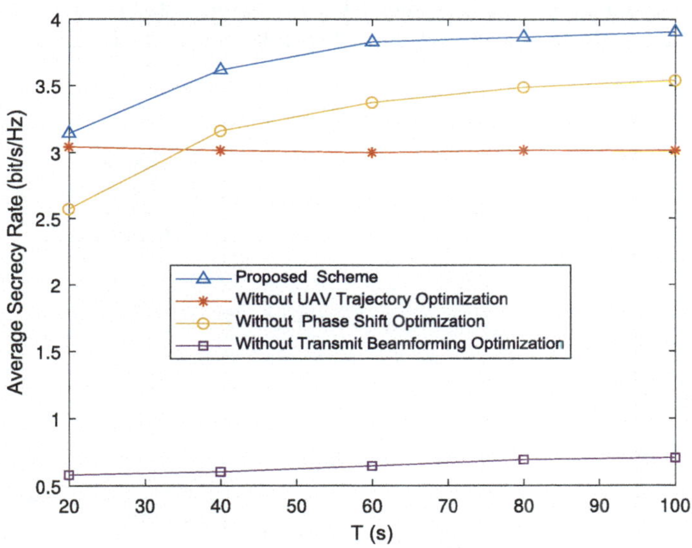


**Fig. 4.3** The secrecy performance comparison of the proposed scheme and other benchmarks

### 4.4.3 Effects of Antennas and Reflecting Elements

In Fig. 4.4, we depict the average secrecy rate versus the number of passive reflecting elements. It is important to note that in this context, we consider the utilization of a square RIS with dimensions represented by $M_x$ and $M_z$, where the total number of reflecting elements is denoted as $M = M_x M_z$. As anticipated, the average secrecy rate is observed to increase with either $M_x$ or $M_z$. This phenomenon can be attributed to the fact that a larger RIS configuration provides greater passive beamforming gain for the legitimate user while avoiding a commensurate increase in eavesdropping power. Upon comparing the curves associated with various numbers of transmit antennas, we can conclude that the augmentation of antenna resources on the UAV also yields improvements in security performance. Therefore, the co-deployment of a substantial number of antennas and reflecting elements boosts both active and passive beamforming, ultimately resulting in a higher secrecy rate.

### 4.4.4 Secrecy Performance Analysis

To gain deeper insights, we compare our proposed Max $\bar{R}_S$ scheme with a legitimate rate maximization scheme named Max $\bar{R}_l$, as illustrated in Figs. 4.5 and 4.6. In Fig. 4.5, we evaluate how the average achievable rates of the legitimate user ($\bar{R}_l$)

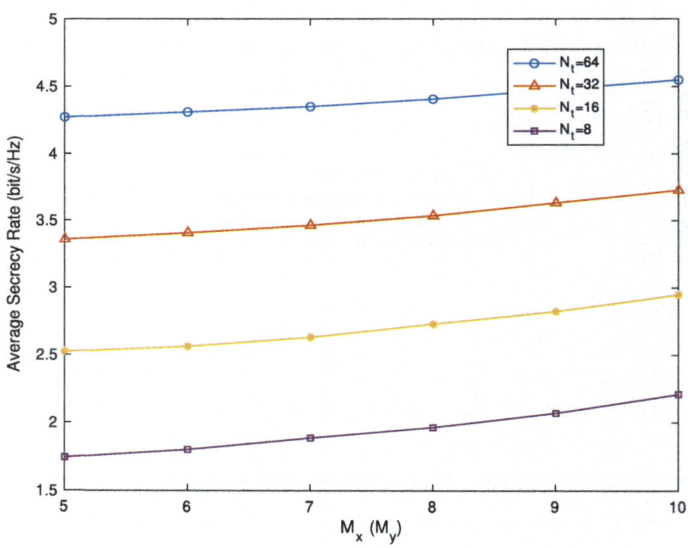


**Fig. 4.4** Average secrecy rate of the proposed scheme versus different numbers of antennas and reflecting elements

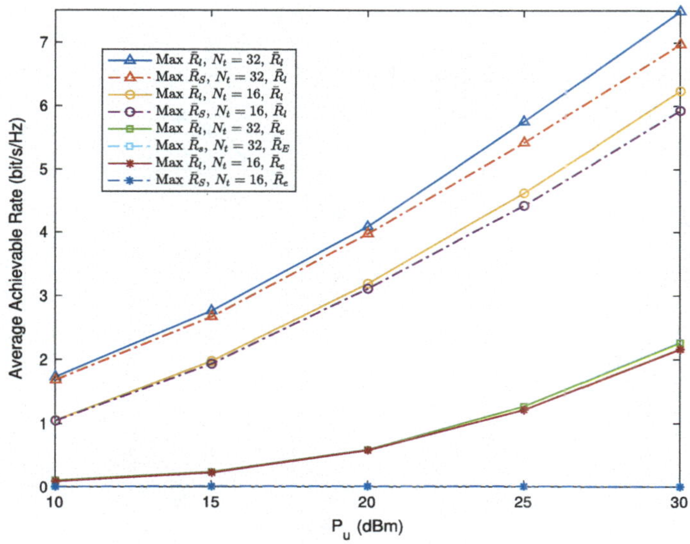


**Fig. 4.5** Achievable rates of legitimate user and eavesdropper versus transmit power

and the eavesdropper ($\bar{R}_e$) vary with transmit power $P_U$. This analysis considers scenarios with $N_t = 16$ and $N_t = 32$. The results show that for both schemes, $\bar{R}_l$ increases as $P_U$ grows. However, there is a significant disparity of $\bar{R}_e$ between the two schemes. In the case of Max $\bar{R}_l$, $\bar{R}_e$ rises with increasing $P_U$, while for Max $\bar{R}_S$, $\bar{R}_e$ remains nearly constant and close to zero, regardless of changes in $P_U$. It can be observed that the legitimate rate maximization scheme consistently yields a slightly higher legitimate rate, with the performance gap widening as $P_U$ increases. Notably, Max $\bar{R}_S$ significantly outperforms Max $\bar{R}_l$ in terms of eavesdropping rate ($\bar{R}_e$). This emphasizes the effectiveness of our joint optimization in countering eavesdropping threats.

In Fig. 4.6, we compare the average secrecy rate $\bar{R}_S$ for both schemes. It can be observed that $\bar{R}_S$ exhibits a rising trend as $P_U$ increases across all cases. Furthermore, our proposed scheme consistently outperforms the legitimate rate maximization scheme, with the performance gap between them widening as $P_U$ grows. Combining the results in Figs. 4.5 and 4.6, it is concluded that although both schemes exhibit similar behavior in terms of increasing $\bar{R}_l$ with rising $P_U$, the crucial difference lies in $\bar{R}_e$. For Max $\bar{R}_S$, $\bar{R}_e$ remains invariant, whereas for Max $\bar{R}_l$, $\bar{R}_e$ increases, as discussed in Fig. 4.5. Therefore, the increase in $\bar{R}_S$ due to higher $P_U$ is more pronounced in our proposed scheme. Additionally, Fig. 4.6 shows that the average secrecy rate experiences a significant improvement when transitioning from 16 to 32 antennas, aligning with the findings in Fig. 4.5.

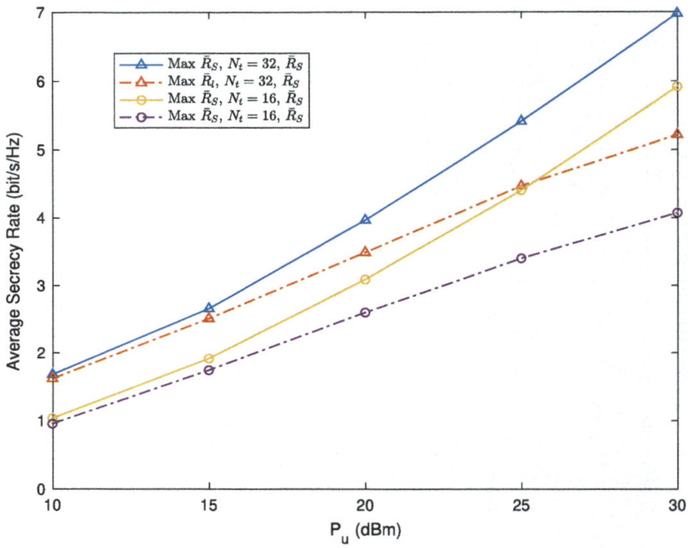


**Fig. 4.6** Comparison of average secrecy rate for different schemes

## 4.5 Conclusions and Open Challenges

This chapter focuses on secure transmission design for RIS-assisted UAV communication networks. Our goal is to leverage the inherent mobility of UAV, along with the beamforming capabilities of antennas and reflecting elements, to perform joint optimization to maximize the average secrecy rate. However, it is worth noting that this joint optimization task poses significant challenges, due to its non-convex and intractable nature. To address this, we break it down into three distinct sub-problems and propose an iterative algorithm for its resolution. To demonstrate the effectiveness of our proposed algorithm, we present simulation results. While the incorporation of RIS technology into UAV networks represents a significant advancement in establishing flexible and secure aerial-ground communications, several challenges remain to be addressed in its practical implementation, as outlined below.

- **Channel Estimation**: Achieving the envisioned performance enhancements by the UAV and RIS integration crucially depends on precise CSI estimation. Imperfect CSI acquisition can compromise beamforming gains, yet higher estimation accuracy typically leads to elevated training overhead and power consumption, especially with large RIS configurations. Therefore, future research should enhance channel estimation accuracy without incurring excessive training overhead or power consumption.
- **Robust Physical Layer Security**: Our chapter has assumed perfect eavesdropping CSI availability in the considered network. However, designing secure systems at the physical layer that involves imperfect eavesdropping CSI is more robust and

practical. Beyond beamforming, cooperative jamming offers an effective method of degrading eavesdropping channel quality. Nonetheless, UAVs are susceptible to detection via visual or radar scans, making them vulnerable to jamming attacks. Consequently, future efforts should explore collaborative secure transmission techniques to enhance security when integrating UAV and RIS.

- **Energy Efficiency**: While an RIS typically requires only limited and static power due to the absence of power amplifiers, the energy-saving aspect of UAVs is a crucial concern. UAVs suffer from their onboard energy limitations and notably higher propulsion energy consumption, which constrains their endurance and overall performance. Moreover, the UAV's endurance is subject to various external factors, such as load, flight speed, altitude, environmental conditions, and flight actions. Therefore, future investigations should focus on designing energy-efficient trajectory and beamforming strategies for RIS-assisted UAV communications without compromising performance.

## References

1. Zeng Y, Zhang R, Lim TJ (2016) Wireless communications with unmanned aerial vehicles: opportunities and challenges. IEEE Commun Mag 54(5):36–42
2. Wu Q, Zhang R (2019) Intelligent reflecting surface enhanced wireless network via joint active and passive beamforming. IEEE Trans Wirel Commun 18(11):5394–5409
3. Jiang X, Sheng M, Zhao N et al (2022) Green UAV communications for 6G: a survey. Chin J Aeronaut 35(9):19–34
4. Pang X, Tang J, Zhao N, Zhang X, Qian Y (2021) Energy-efficient design for mmWave-enabled NOMA-UAV networks. Sci China Inf Sci 64:1–4
5. Al-Hourani A, Kandeepan S, Lardner S (2014) Optimal LAP altitude for maximum coverage. IEEE Wirel Commun Lett 3(6):569–72
6. Mozaffari M, Saad W, Bennis M, Nam YH, Debbah M (2019) A tutorial on UAVs for wireless networks: applications, challenges, and open problems. IEEE Commun Surv Tutor 21(3):2334–60
7. Chen X, Li D, Yang Z, Chen Y, Zhao N, Ding Z, Yu FR (2020) Securing aerial-ground transmission for NOMA-UAV networks. IEEE Netw 34(6):171–177
8. Wu Q, Zhang R (2019) Towards smart and reconfigurable environment: intelligent reflecting surface aided wireless network. IEEE Commun Mag 58(1):106–112
9. Wu Q, Zhang S, Zheng B, You C, Zhang R (2021) Intelligent reflecting surface-aided wireless communications: a tutorial. IEEE Trans Commun 69(5):3313–3351
10. Alfattani S, Jaafar W, Hmamouche Y et al (2021) Aerial platforms with reconfigurable smart surfaces for 5G and beyond. IEEE Commun Mag 59(1):96–102
11. Pang X, Sheng M, Zhao N, Tang J, Niyato D, Wong KK (2021) When UAV meets IRS: expanding air-ground networks via passive reflection. IEEE Wirel Commun 28(5):164–170
12. Li S, Duo B, Yuan X, Liang YC, Di Renzo M (2020) Reconfigurable intelligent surface assisted UAV communication: joint trajectory design and passive beamforming. IEEE Wirel Commun Lett 9(5):716–720
13. Ge L, Dong P, Zhang H, Wang JB, You X (2020) Joint beamforming and trajectory optimization for intelligent reflecting surfaces-assisted UAV communications. IEEE Access 8:78702–78712
14. Su Y, Pang X, Chen S, Jiang X, Zhao N, Yu FR (2022) Spectrum and energy efficiency optimization in IRS-assisted UAV networks. IEEE Trans Commun 70(10):6489–6502

15. Hu C, Dai L, Han S, Wang X (2021) Two-timescale channel estimation for reconfigurable intelligent surface aided wireless communications. IEEE Trans Commun 69(11):7736–7747
16. Pang X, Zhao N, Tang J, Wu C, Niyato D, Wong KK (2021) IRS-assisted secure UAV transmission via joint trajectory and beamforming design. IEEE Trans Commun 70(2):1140–1152

# Chapter 5
# Secrecy Analysis of UAV-Assisted RIS-Enhanced Uplink NOMA Systems

**Khawaja Muhammad Hamza, Sarah Basharat, Haejoon Jung, Mikael Gidlund, and Syed Ali Hassan**

**Abstract** In this paper, we consider an Unmanned aerial vehicle (UAV)-assisted reconfigurable intelligent surface (RIS)-enhanced uplink non-orthogonal multiple access (NOMA) communication system, where users (UEs) want to transmit confidential information to the UAV in the presence of an eavesdropper. We evaluate the impact of transmit power on secrecy rate with continuous and discrete phase shifts at RIS under Rayleigh fading channels. The simulation results reveal that the secrecy rate increases with the increase in the transmit power of UEs and discrete phase shift at RIS degrades the system performance.

**Keywords** Unmanned aerial vehicle (UAV) · Reconfigurable intelligent surface (RIS) · Non-orthogonal multiple access (NOMA) · Split ratio · Secrecy rate

---

K. M. Hamza (✉) · S. Basharat · S. A. Hassan
School of Electrical Engineering and Computer Science (SEECS), National University of Sciences & Technology (NUST), Islamabad, Pakistan
e-mail: khamza.msee21seecs@seecs.edu.pk

S. Basharat
e-mail: sarah.phdee21seecs@seecs.edu.pk

S. A. Hassan
e-mail: ali.hassan@seecs.edu.pk

H. Jung
Department of Electronic Engineering, Kyung Hee University, Yongin, South Korea
e-mail: haejoonjung@khu.ac.kr

M. Gidlund
Department of Computer and Electrical Engineering, Mid Sweden University, Sundsvall, Sweden
e-mail: mikael.gidlund@miun.se

© The Author(s), under exclusive license to Springer Nature Singapore Pte Ltd. 2026 
I. Ahmad et al. (eds.), *Reconfigurable Intelligent Surfaces for Unmanned Aerial Vehicle Communication*, Unmanned System Technologies
https://doi.org/10.1007/978-981-96-9662-8_5

## 5.1 Introduction

Sixth-generation (6G) communications are made possible by three-dimensional (3D) broadband connectivity by using heterogeneous networks utilizing high-altitude platforms (HAP) and unmanned aerial vehicles (UAVs) [1, 2]. Fast deployment, extensive coverage, and affordable communication mechanisms are offered by HAP and on-demand deployment of UAVs hovering at low altitudes. These methods are especially well suited for servicing remote areas and catastrophe locations when terrestrial communications are not accessible. Free space optical (FSO) transmission is commonly utilized to support heterogeneous networks, including HAP and UAV, because of its benefits in terms of ultra-high data speeds, license-free spectrum, and high security of FSO. It enables ground users to send messages to the UAV through the RF link, where the electrical signal is converted to an optical signal and thereafter sent over the FSO link to the HAP [3].

On the other hand, non-orthogonal multiple access (NOMA) will be essential in sixth-generation (6G) systems [4]. More user connections and spectral efficiency may be achieved with NOMA than with conventional orthogonal multiple access (OMA) [5]. Due to broadcasting nature of signals, the transfer of information between UAV and legitimate user is critical.

Physical layer security has received a lot of attention as a means of ensuring secure wireless communications and preventing unwanted users from listening in [6, 7]. Often, application layer implementation of classical cryptographic algorithms is used to achieve secure communication. When establishing secure communication channels, however, the nature of wireless transmission poses a number of challenges for key distribution and administration [8]. To increase the security of wireless communication, physical layer security has been developed using ideas from information theory. Recently, the research community has been impacted by secret communication for multiantenna wiretap channels because it offers additional degrees of freedom and diversity benefits [9].

Reconfigurable intelligent surface (RIS) has been presented as a novel, low-cost technique to save transmit power and maximize energy-saving measures [10, 11]. RIS usually comprises a large multitude of inexpensive, passive reflecting components that the phase-shifted incident signal that varies. Consequently, the RIS and NOMA together might be an effective way to improve the privacy performance of the wireless network.

RIS-assisted NOMA systems create a secure signal zone for the intended user by coherent phase shifting from RIS elements, resulting in signal attenuation for unauthorized users, which eventually improves the secrecy performance. In this regard, the authors in [12] proposed a novel design of RIS to enhance the PLS in the RIS-aided NOMA network. In [13], the authors investigated the impact of residual hardware impairment on the Internet-of-Things (IoT) secrecy performance of RIS-assisted NOMA networks. In [14], the authors investigated the network secrecy performance of the RIS-assisted various emerging communication techniques, such as unmanned aerial vehicles (UAVs), NOMA, millimeter wave (mmWave), and terahertz (THz)

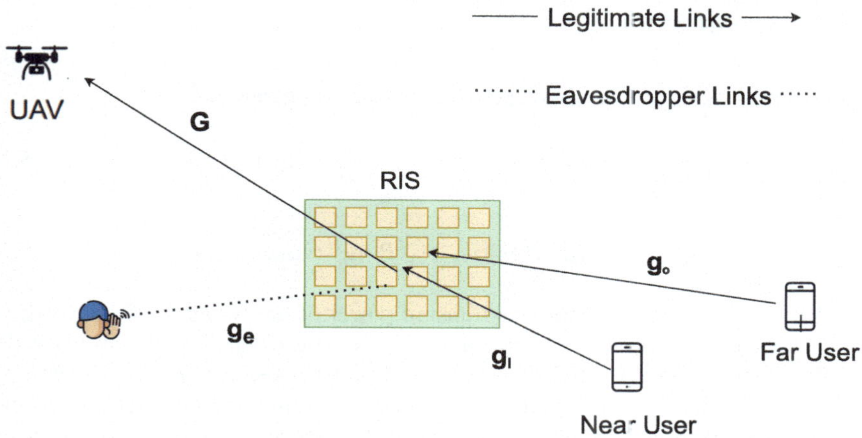


**Fig. 5.1** An illustration of secure UAV-assisted RIS-enhanced uplink NOMA communication system

communications. In [15], the authors investigated the secrecy performance of the RIS-aided NOMA network for two-user NOMA for both internal and external eavesdropping. Different from previous works, the main contribution of this work can be summarized as follows:

- We consider an UAV-assisted RIS-enhanced uplink NOMA communication system, where UEs intend to transmit confidential information with the aid of an RIS to the UAV in the presence of an eavesdropper.
- We analyze the secrecy rate of the considered system under Rayleigh fading channels. Moreover, our simulation results demonstrate the impact of transmit power on the secrecy rate with continuous and discrete phase shifts at RIS elements.

## 5.2 System Model

As illustrated in Fig. 5.1, we consider an RIS-assisted uplink NOMA network, where two single antenna users, denoted by N and F, intend to deliver confidential information, $x_N$ and $x_F$, respectively, to a single antenna aerial BS in the presence of an eavesdropper with the aid of $K$ passive reconfigurable elements of an RIS, denoted by $\mathcal{K} = \{1, 2, \ldots, K\}$. More specifically, N denotes the NOMA near user, while F denotes the NOMA far user.

### 5.2.1 Signal Model

The received signals at the UAV and eavesdropper are given by

$$y_u = \sqrt{P_N}(\mathbf{G}^H \Theta \mathbf{g}_N)x_N + \sqrt{P_F}(\mathbf{G}^H \Theta \mathbf{g}_F)x_F + n_u, \tag{5.1}$$

and

$$y_e = \sqrt{P_N}(\mathbf{g}_e^H \Theta \mathbf{g}_N)x_N + \sqrt{P_F}(\mathbf{g}_e^H \Theta \mathbf{g}_F)x_F + n_e, \tag{5.2}$$

respectively, $P_N$ and $P_F$ denote the transmit power of the N and F, respectively, $\mathbf{G} \in \mathbb{C}^{K\times 1}$, $\mathbf{g}_N \in \mathbb{C}^{N\times 1}$, $\mathbf{g}_F \in \mathbb{C}^{K\times 1}$, and $\mathbf{g}_e \in \mathbb{C}^{N\times 1}$ are the channel vectors from the BS to RIS, RIS to N, RIS to F, and RIS to the eavesdropper, respectively, $n_u \sim \mathcal{CN}(0, \sigma_u^2)$ and $n_e \sim \mathcal{CN}\left(0, \sigma_e^2\right)$ are the additive white Gaussian noise (AWGN) at the UAV and eavesdropper, respectively, with zero mean and variances $\sigma_u^2$ and $\sigma_e^2$, respectively, and $\mathbf{G}^H$ denotes the Hermitian transpose of $\mathbf{G}$. Similarly in (1) and (2), $\Theta$ = diag $\left(e^{j\theta_1}, e^{j\theta_2}, \ldots, e^{j\theta_K}\right) \in \mathbb{C}^{K\times K}$ is a diagonal matrix, and $\theta_k \in \mathcal{F}_i$ denotes the phase shift of the $k$-th reflecting element of the RIS, where $\mathcal{F}_i$ is the feasible set of phase shift. We consider two feasible sets of phase shifts $\mathcal{F}_1$ and $\mathcal{F}_2$, where $\mathcal{F}_1 = [0, 2\pi)$ is the set of *continuous phase shifts*, and $\mathcal{F}_2 = \{0, \Delta\theta, \ldots, (L-1)\Delta\theta\}$ is the set of *discrete phase shifts*, obtained by uniformly quantizing the interval $[0, 2\pi)$, where $\Delta\theta = 2\pi/L$, and $L = 2^b$ denotes the discrete phase shift levels which have a $b$-bit resolution.

### 5.2.2 Channel Model

Without the loss of generality, all channels are characterized by distance-dependent path loss and small-scale fading. The distance-dependent path loss is modeled as $L(d_x) = d_x^{-\alpha}$ where $x = \{\mathbf{G}, \mathbf{g}_N, \mathbf{g}_F, \mathbf{g}_e\}$, $d_x$ and $\alpha$ denotes the distance and path loss exponent of the corresponding links, respectively. In order to incorporate the small-scale fading we assume Rayleigh fading.

### 5.2.3 SINR, Achievable Rate, and Secrecy Rate

Based on the conventional uplink NOMA scheme, the BS first decodes the signal of strong user in the NOMA pair. In this work, we assume that N is stronger than the F. Thus, the received signal-to-interference-plus-noise ratio (SINR) at the BS and eavesdropper to decode $x_N$ can be expressed as

$$\gamma_N = \frac{P_N \left(\sum_{k=1}^{K} |G_k| \left|g_{N,k}\right|\right)^2}{P_F \left(\sum_{k=1}^{K} |G_k| \left|g_{F,k}\right|\right)^2 + \sigma_u^2}, \tag{5.3}$$

and

$$\gamma_{e,N} = \frac{P_N \left|\sum_{k=1}^{K} G_k g_{e,k} e^{j\theta_k}\right|^2}{P_f \left|\sum_{k=1}^{K} G_k g_{e,k} e^{j\theta_k}\right|^2 + \sigma_e^2}, \tag{5.4}$$

respectively, where $G_k$, $g_{N,k}$, $g_{F,k}$, and $g_{e,k}$ denote the $k$-th element of $\mathbf{G}$, $\mathbf{g}_N$, $\mathbf{g}_F$, and $\mathbf{g}_e$, respectively. Then, the UAV and eavesdropper applies successive interference cancellation (SIC) to remove $x_N$ and decode $x_F$. The received signal-to-noise ratio (SNR) at the UAV and eavesdropper to decode $x_N$ can be expressed as

$$\gamma_F = \frac{P_F \left(\sum_{k=1}^{K} |G_k| \left|g_{F,k}\right|\right)^2}{\sigma_u^2}, \tag{5.5}$$

and

$$\gamma_{e,F} = \frac{P_F \left|\sum_{k=1}^{K} g_{e,k} g_{2,k} e^{j\theta_k}\right|^2}{\sigma_e^2}, \tag{5.6}$$

respectively. The corresponding achievable rates are given by

$$R_N = \log_2 \left(1 + \frac{P_N \left(\sum_{k=1}^{K} |G_k| \left|g_{N,k}\right|\right)^2}{P_F \left(\sum_{k=1}^{K} |G_k| \left|g_{2,k}\right|\right)^2 + \sigma_u^2}\right), \tag{5.7}$$

$$R_F = \log_2 \left(1 + \frac{\frac{P_F}{(d_G d_{g_F})^{\alpha}} \left(\sum_{k=1}^{K} |G_k| \left|g_{2,k}\right|\right)^2}{\sigma_u^2}\right), \tag{5.8}$$

$$R_{e,N} = \log_2 \left(1 + \frac{P_N \left|\sum_{k=1}^{K} g_{e,k} g_{1,k} e^{j\theta_k}\right|^2}{P_F \left|\sum_{k=1}^{K} g_{e,k} g_{2,k} e^{j\theta_k}\right|^2 + \sigma_e^2}\right), \tag{5.9}$$

,

$$R_{e,F} = \log_2 \left(1 + \frac{P_F \left|\sum_{k=1}^{K} g_{e,k} g_{F,k} e^{j\theta_k}\right|^2}{\sigma_e^2}\right). \tag{5.10}$$

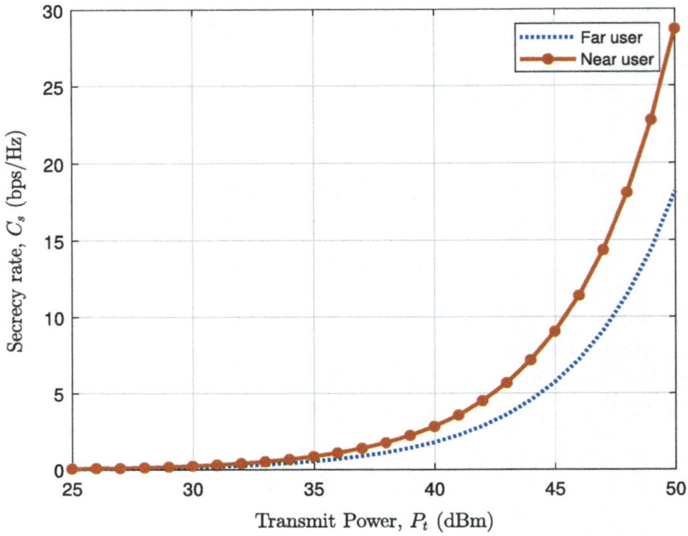


**Fig. 5.2** Secrecy rate for varying transmit power of users

The secrecy rate for transmitting confidential information to the UAV in the presence of an eavesdropper can be expressed as [14, 15]

$$C_{\rho_s} = \left[R_\rho - R_{e,\rho}\right]^+ = \left[\log_2\left(1+\gamma_\rho\right) - \log_2\left(1+\gamma_{e,\rho}\right)\right]^+, \tag{5.11}$$

where $[z]^+ \triangleq \max(0, z)$ and $\rho \in \{N, F\}$.

### 5.2.4 RIS Configuration and Element Splitting

We assume that $K$ RIS elements are split [13] between two users, i.e., a total of $K_F$ elements coherently combine the signal of F, based on that we can define the split factor $\gamma$ as

$$\gamma = \frac{K_F}{K}, \tag{5.12}$$

while a total of $K_N$ elements coherently combine the signal of N, which is given by

$$K_N = K - [\gamma K]. \tag{5.13}$$

Therefore, the RIS-assisted channel gains can be written as

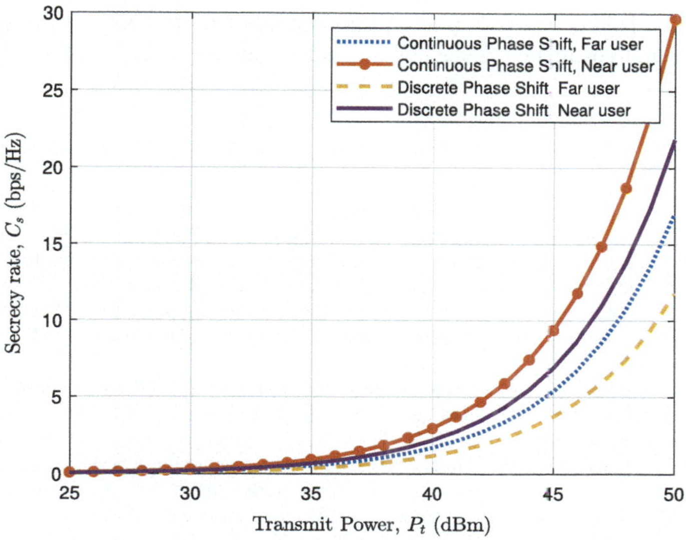


**Fig. 5.3** Secrecy rate for varying transmit power under continuous and discrete phase shifts

$$\left|\mathbf{G}^H \Theta \mathbf{g}_\rho\right|^2 = \underbrace{\left(\sum_{k=1}^{K_\rho} |G_k| \left|g_{\rho,k}\right|\right)^2}_{\text{Coherently Combined}} + \underbrace{\left|\sum_{k=K_\rho+1}^{K} G_k g_{\rho,k} e^{j\theta_k}\right|^2}_{\text{Randomly Combined}}. \quad (5.14)$$

The set of elements which are not configured for the $\rho$-th user will result in the random combining.

Figure 5.2 illustrates the impact of transmit power on near and far users. First, it can observed that with increase in the transmit power the secrecy rate of both users increase. Second, the secrecy rate of the near user will be greater than the far user.

Figure 5.3 illustrates the impact of transmit power on near and far user under continuous and discrete phase shifts of RIS. It can be observed that the secrecy rate for the user with continuous phase shift at the RIS is greater than the user with discrete phase shift at RIS.

## 5.3 Conclusion

This paper investigated the secrecy performance of UAV-assisted RIS- aided uplink NOMA communication system. Our simulation result demonstrated the impact of transmit power with continuous and discrete phase shifts at RIS. Our results revealed that with the discrete phase shift at RIS the secrecy rate of the system degrades. In future, we aim to evaluate the secrecy outage probability of the considered system.

## References


1. Khan SK, Naseem U, Sattar A, Waheed N, Mir A, Qazi A, Ismail M (2020) UAV-aided 5G network in suburban, urban, dense urban, and high-rise urban environments. In: 2020 IEEE 19th international symposium on network computing and applications (NCA), pp 1–4
2. ur Rahman S, Kim G-H, Cho Y-Z, Khan A (2018) Positioning of UAVs for throughput maximization in software-defined disaster area UAV communication networks. J Commun Netw 20(5):452–463
3. Najafi M, Ajam H, Jamali V, Diamantoulakis PD, Karagiannidis GK, Schober R (2020) Statistical modeling of the FSO fronthaul channel for UAV-based communications. IEEE Trans Commun 68(6):3720–3736
4. Nonaka N, Kishiyama Y, Higuchi K (2014) Non-orthogonal multiple access using intra-beam superposition coding and SIC in base station cooperative MIMO cellular downlink. In: 2014 IEEE 80th vehicular technology conference (VTC2014-Fall), pp 1–5
5. Dai L, Wang B, Yuan Y, Han S, Chih-lin I, Wang Z (2015) Non-orthogonal multiple access for 5G: solutions, challenges, opportunities, and future research trends. IEEE Commun Mag 53(9):74–81
6. Tang X, Liu R, Spasojevic P, Poor HV (2008) Interference-assisted secret communication. In: IEEE Information theory workshop, pp 164–168
7. Zhang R, Song L, Han Z, Jiao B, Debbah M (2010) Physical layer security for two way relay communications with friendly jammers. In: 2010 IEEE global telecommunications conference GLOBECOM 2010, pp 1–6
8. Hanif A, Badrudduza ASM, Hossen MS, Kundu MK, Sarkar MZI (2019) Secure outage performance analysis of multicasting through $\alpha$—$\mu$ fading channel. In: 2019 3rd international conference on electrical, computer & telecommunication engineering (ICECTE), pp 65–68
9. Khisti A, Wornell GW (2010) Secure transmission with multiple antennas—Part II: the MIMOME wiretap channel. IEEE Trans Inf Theory 56(11):5515–5532
10. Shen H, Xu W, Gong S, He Z, Zhao C (2019) Secrecy rate maximization for intelligent reflecting surface assisted multi-antenna communications. IEEE Commun Lett 23(9):1488–1492
11. Shu F, Teng Y, Li J, Huang M, Shi W, Li J, Wu Y, Wang J (2021) Enhanced secrecy rate maximization for directional modulation networks via IRS. IEEE Trans Commun 69(12):8388–8401
12. Tahir B, Schwarz S, Rupp M (2021) Outage analysis of uplink IRS-assisted NOMA under elements splitting. In: 2021 IEEE 93rd vehicular technology conference (VTC2021-spring), pp 1–5
13. Basharat S, Ali Hassan S, Pervaiz H, Mahmood A, Ding Z, Gidlund M (2021) Reconfigurable intelligent surfaces: potentials, applications, and challenges for 6G wireless networks. IEEE Wirel Commun 28(6):184–191
14. Basharat S, Pervaiz H, Hassan SA, Ansari RI, Jung H, Dev K, Huang G (2022) Intelligent radio resource management in reconfigurable IRS-enabled NOMA networks. Phys Commun 53:101744
15. Tang Z, Hou T, Liu Y, Zhang J, Zhong C (2021) A novel design of RIS for enhancing the physical layer security for RIS-aided NOMA networks. IEEE Wirel Commun Lett 10(11):2398–2401

# Chapter 6
# Energy-Efficient Resource Allocation in RIS-Assisted UAV Communication

**Muhammad Ammad Awan, Muhammad Kamran, and Ayaz Ahmad**

**Abstract** The use of unmanned aerial vehicles (UAVs) as flying base station in wireless communication has attracted a lot of interest in research and industry. UAV as a base station has more coverage as compared to the ground base station and has a direct or line-of-sight (LOS) communication link with most of the ground users. In such scenarios, propagation losses are minimum and the users are served with the least possible amount of transmission power. However, it may sometimes occur that there is no LOS communication link, i.e., an obstacle comes between the UAV and the ground user causing an invisibility. This results in more propagation losses and high transmission power is needed for communication between UAV and ground users. Fortunately, in no LOS link cases, the transmission power can be minimized by using reconfigurable intelligent surfaces (RIS). These surfaces have the ability to reconfigure the reflection of the intended signal coming from the UAV and send it to the desired users by adjusting its direction and phase. The direct mode of communication (i.e., from the UAV to ground users) and the reflecting mode of communication (i.e., from UAV to RIS and RIS to ground users) consume different amounts of power, and one between the two should be selected intelligently to capitalize on the use of RIS joint with UAV. In this chapter, we consider a wireless communication model with an RIS-assisted UAV base station to serve the multiple ground users. We consider the problem of communication mode selection and power allocation with objective of maximizing the bits per joule to maximize the system energy efficiency. For this purpose, we formulate a mathematical problem and find a solution to that problem. Then, with the help of numerical simulations, we evaluate the energy efficiency performance of our proposed solution.

---

M. A. Awan · M. Kamran
Department of Electrical and Computer Engineering, COMSATS University Islamabad, Wah Campus, Pakistan
e-mail: mammadawan7@gmail.com

M. Kamran
e-mail: dr_m_kamran@ciitwah.edu.pk

A. Ahmad (✉)
Department of Electrical Engineering, College of Engineering, King Faisal University, Al-Ahsa, Saudi Arabia
e-mail: ayahmad@kfu.edu.sa

© The Author(s), under exclusive license to Springer Nature Singapore Pte Ltd. 2026

I. Ahmad et al. (eds.), *Reconfigurable Intelligent Surfaces for Unmanned Aerial Vehicle Communication*, Unmanned System Technologies,
https://doi.org/10.1007/978-981-96-9662-8_6

**Keywords** Resource allocation · UAV-assisted communication · Intelligent reflecting surfaces

## 6.1 Introduction

Distributed network intelligence in next-generation networks will enable advanced mobile networking and spectrum co-existence among an increasing number of passive and active radio services. Wireless networks are designed to connect a rapidly growing number of users with complex demands. Upcoming wireless networks will combine flight and ground networks to achieve high connectivity [1]. UAVs have gained the interest of manufacturing and academia in recent years due to their great mobility and low manufacturing costs. Unmanned Aerial Vehicles (UAVs) are a type of aircraft that can be flown autonomously with the use of software embedded in their system, or they can be flown remotely. Flight mode and navigation are the two essential UAV functions. UAVs require a power source, such as a battery or fuel, in order to fly. Depending on the situation, built-in circuits or control centers may be used to operate them. It is known as a remotely piloted vehicle (RPV) when it is operated remotely from the ground, and it needs dependable wireless connectivity to operate. A specific control system may be needed for large UAVs, and these systems can be mounted aboard vehicles or in vehicles to allow proximity to UAVs with restricted range. UAVs are primed and ready to be the 5G game changers and success factors of upcoming connectivity. UAVs have unique qualities that include comfortable stay and path mobility, affordable installation expenses, and the skill to form uninterrupted line-of-sight (LOS) pathways, among other hubs. As a result, in addition to using the data connection, drones are seen as carrier service units that help both past and present mobile network users function better.

UAVs can be viewed as the promising method of communication in areas where ground communications are effected by natural disasters or in any environment that is insecure [2]. Because they have a better line-of-sight (LOS) connection to users on the ground, UAVs can be moved easily and quickly when needed. They also have a dynamic height that allows them to meet quality-of-service (QOS) requirements that depend on the number of users, the necessary data rate, and interference effects. There are several uses for UAVs in wireless networks. When used in combination with terrestrial base stations, UAVs improve coverage, quality and reliability, and energy efficiency. Within cellular networks, they can potentially be utilized as flying mobile terminals.

In any region that lacks security or where natural disasters impact ground communications, unmanned aerial vehicles (UAVs) hold great promise as a means of communication. Because they have a better line-of-sight (LOS) connection to users on the ground, UAVs can be moved easily and quickly when needed. They also have a dynamic height that allows them to meet quality-of-service (QOS) requirements that depend on the number of users, the necessary data rate, and interference effects. There are several uses for UAVs in wireless networks. When used in combination

**Fig. 6.1** Different uses of UAVs

with terrestrial base stations, UAVs improve coverage, quality, reliability, and energy efficiency. Within cellular networks, they can potentially be used as flying mobile terminals.

The mobility of UAVs also creates new opportunities for performance enhancement by enabling the UAV state to be dynamically modified to best suit the communication environment. To improve communication performance even more, adaptive communications can be used with UAV movement control. For instance, when a UAV has a strong connection to the terrestrial base terminals, it may slow down to hold the strong wireless connection and send extra data to the base terminals in addition to broadcasting at higher rates. UAV-assisted wireless communication is a significant component of the upcoming wireless systems, which will need to support a wider range of applications while providing orders-of-magnitude capacity improvements over current systems, thanks to these obvious benefits [3]. Drones can be used to replace any existing communication infrastructure to provide seamless wireless coverage throughout the service area. Timely recovery from partial or complete damage to infrastructure caused by natural disasters and the removal of base stations in crowded areas, such as a stadium during a sporting event, are examples of these situations. UAV-assisted relay, where UAVs are used to provide wireless connectivity between different users. UAV-assisted relay, where the UAV is used to provide wireless connectivity between different users or groups of users that do not have a reliable LOS communication link, such as the frontline and control center for emergency response (Fig. 6.1).

The term "reconfigurable intelligent surface" (RIS) refers to a new technology that is also known as "intelligent reflecting surface" or "software-controlled surface." A vast number of reflecting devices make up the two-dimensional meta-surface known as RIS, whose phase and amplitude can be modified and manipulated in the wireless environment to direct incoming signals in the required directions [4]. They aid in a variety of goals, including as improving the user-side signal or eliminating the impact of interference, because of their adaptive nature to intelligently manage the incoming signals. A result of their low cost and nearly passive design, RISs are regarded as the most cost-effective alternative for networks beyond 5G and 6G networks. The main issue with next-generation networks is resolved by the passive nature of RIS, which reduces energy consumption. They are excellent choices to fit in the next-generation networks because of their ease of implementation. To enhance their use, they can be mounted on any shape, including rooftops and unmanned aerial vehicles (UAVs).

In particular at higher frequencies, the RIS-aided UAV networking design shows great promise for boosting wireless communications and managing the growing complexity of the wireless channel. ABS (Aerial Base Station) and AR (Aerial Relay) technology, which are seen as crucial enablers for dynamic and adaptive coverage, will be very helpful for future cellular network installations. However, the energy required for these active aerial communications is massive [5]. Compared to ABS and AR, the RIS is a more passive choice. By deploying the RIS-UAV, throughput and capacity can be increased. Half-duplex (HD) mode, which is less effective than full-duplex, is more likely to be used when utilizing an AR (FD). A major advantage over active frequency division (FD) relay systems is that the RIS relay process is resistant to amplification and self-interference caused by antenna noise.

When it comes to deployment, the architecture of RIS-aided UAVs presents considerable difficulties. They face a variety of difficulties, including RIS controller overhead, channel modeling, and channel estimation. For the development of RIS-UAV channels, many factors must be taken into account [6]. Because of the barriers and distance of an edge user, the base station cannot give total coverage. UAV-based RIS will be deployed in the area, providing better coverage by using the RIS to beam-form the signal. Additionally, RIS helps suppress unwanted signals and interference from nearby base stations. RIS is useful for boosting or canceling the signal, especially for edge users. UAVs continue to improve their capabilities as mobile air nodes.

### *6.1.1 Motivation*

The motivation behind this research is to address the challenges of providing reliable wireless communication in scenarios where direct line-of-sight links are not available. In such scenarios, conventional communication systems suffer from high propagation losses, which leads to poor signal quality and low data rates. The use of unmanned aerial vehicles (UAVs) as flying base stations is a promising solution to overcome these challenges, as they can be deployed quickly and provide wider coverage. However, even UAVs have limitations when it comes to providing reliable communication in non-line-of-sight scenarios. To overcome this limitation, researchers have proposed the use of reconfigurable intelligent surfaces (RIS), which are essentially planar structures that can reflect incident signals in a controllable manner. By placing an RIS in the environment, the reflected signals can be manipulated in a way that enhances the signal quality and provides a direct link between the UAV and the users on the ground.

### 6.1.2 Contribution

The contribution of this research is the development of a model that maximizes energy efficiency by maximizing the number of bits transmitted per joule of energy. The proposed model involves a UAV, an RIS, and multiple ground users, with two transmission modes: direct transmission and RIS-assisted transmission. The suggested method maximizes efficiency by judiciously deciding which communication mode to use and how much power to provide it. In addition, the authors describe a Dinkelbach approach for decomposing the non-convex, non-linear fractional programming problem into two manageable sections. The study not only builds a model of the system and an algorithm to run it, but also conducts an analysis of the suggested method's efficiency using simulated data. The simulation results show that the suggested method is superior to the baseline methods in terms of energy efficiency and system performance. Additionally, the effect of several system factors on the performance of the system is evaluated, including the number of users, the distance between the UAV and the ground users, and the number of RIS pieces. The results of this investigation may be used to improve the performance of communication systems that use UAV-RIS technology.

### 6.1.3 Objective

The objective of this chapter is to develop an energy-efficient communication system using UAVs and RISs that can provide reliable and high-quality communication services to ground users. The proposed approach aims to maximize the energy efficiency of the system while maintaining the quality-of-service (QoS) requirements of the users. This research is essential for the development of next-generation wireless communication systems, particularly in scenarios where traditional communication infrastructure is not available or feasible, such as disaster relief operations, remote areas, and military applications.

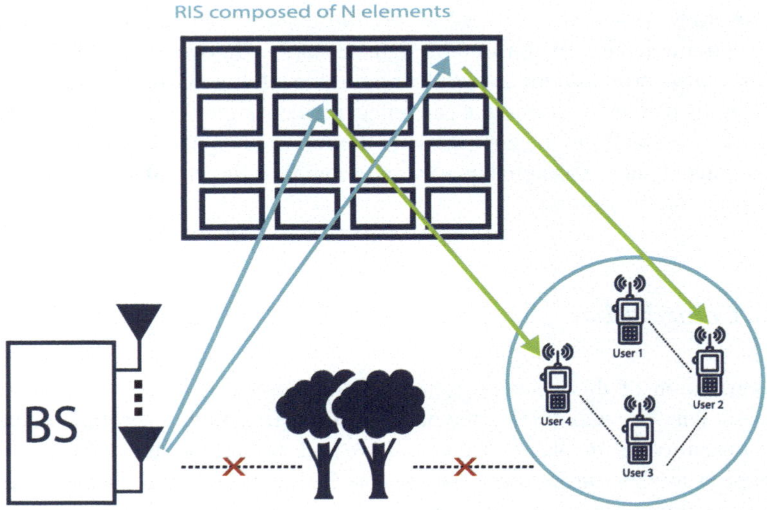


**Fig. 6.2** Configuration of RIS

## 6.2 Literature Review

There are numerous articles on how to optimize different settings to enhance a UAV's performance. The UAVs entire performance is influenced by its weight, energy usage, and trajectory. By adjusting these settings, system performance can be significantly improved. To improve transmission between the point of origin and the point of arrival, RIS's phase shift is equally important. Therefore, it is crucial to identify the precise parameters that help to improve and optimize the system's performance. So we go through a number of methods and characteristics that might be applied to enhance the effectiveness of the system. Changing a few parameters, whether on the RIS or the UAV, can improve the system's performance. It may just be necessary to utilize a UAV to assist in the transmission of radio signals, but studies show that the position, trajectory, and movement of the UAV, when taken into account, significantly increase the efficiency compared to traditional approaches [7].

Any system's effectiveness is crucial since it determines whether or not the system can be implemented. We have considered a UAV-assisted RIS communication network that takes into account bandwidth and energy. In specific situations, optimizing beam-forming vectors at the ground station, RIS, or drone has been the subject of numerous papers. These articles greatly enhanced performances, particularly in terms of energy efficiency and other factors. Therefore, one possible future line of research is to investigate the effects of enhancing beam-forming parameters in UAV-assisted IRS in different circumstances [8].

Machine learning has experienced fast growth since its conception as a result of its capacity to autonomously adjust to changing environmental conditions by learning from huge volume of data [9]. Researchers investigated the potential applications of deep reinforcement learning (DRL) and federated learning to various kinds of unmanned aerial vehicles (UAVs) RIS networks. Only a small handful of articles have considered coupled UAV-RIS systems. An enabling technology to improve radio signal propagation in UAV wireless networks is the RIS. RIS is able to dynamically change wireless channels to improve throughput by carefully adjusting the signal reflection via a significant number of inexpensive passive reflecting devices. Therefore, it is believed that the newly designed UAV-assisted wireless network supported by the RIS, which consists of both active and passive components, will be incredibly convincing in achieving a long-term profitable growth.

As depicted in Fig. 6.2, it may be especially helpful for systems that operate in high-frequency bands where dead zone deficiencies brought on by blockage and absorption are common and result in breakdowns that otherwise might necessitate the implementation of mega-BS with increased power consumption and backhaul expense. Despite its tremendous ability, RIS must overcome new obstacles before it can be effectively used in UAV-assisted wireless networks. These obstacles include reflection coefficient optimization, channel estimation, implementation, etc. [10].

### *6.2.1 UAV Placement and Resource Allocation*

The literature on flying base stations (FBSs) has grown tremendously in recent years including as quick deployment and flexibility. Authors in [11] created a general mixed-integer non-linear formulation for FBSs that caters the specification of its component parts. This work has carried out a thorough literature review on UAV-BS position optimization in Next-Generation Wireless Networks (NGWNs). This work suggests that novel approaches to enhance energy efficiency should be adopted to tackle the issues with the FBSs such as its short service time due its battery-dependent operating lifecycle. In general, it seems that FBSs' practical application is unlikely to be adopted.

An efficient UAV-BS placement strategy is proposed by authors in [12] which maximizes the number of users covered while the least amount of transmitted power is used. For the scenarios where cellular networks become unavailable due to disasters (such as earthquakes or floods), the authors in [13] proposed a wide structural system by providing cellular coverage for indoor users with a single UAV. They assume that indoor user locations are equally distributed throughout every floor and provide particle swarm optimization approach for locating an effective 3D UAV deployment to lower overall transmit power looked-for internal users.

Authors in [14] showed that in-band full-duplex (IBFD) may increase spectrum efficiency in the context of drone base stations (DBSs). DBSs have attracted interest because of their adaptable, affordable, and wirelessly movable characteristics. When DBSs are equipped with in-band IBFD, which could boost the spectrum capabilities, and enhance the throughput, and thus the number of DBSs required to deliver the same level of coverage could be decreased [15].

Resource and power distribution joint with efficient 3D UAV placement are highly challenging. The authors in [18] propose that to enhance user coverage in uplink transmission, the 3D position and attenuation issue should be optimized simultaneously. This work is carried out for multiple UAV deployments in suburban setting which shows the air-to-ground channel characteristics, and SINR criteria affect aerial base station (ABS) coverage probability for different ABS heights. SINR criteria and ABS coverage by changing ABS heights are always a trade-off. For three-dimensional deployment of UAVs, research work in [19] had introduced new learning-based strategies to help terrestrial networks in downlink for crowded occasions. They utilized the multi-armed savage and comfort strategies to learn the placement of drones, and then apply reinforcement learning tools to complete the drone placement.

A deep Q network trained version was suggested by the authors in [23] for the optimal placement of UAVs as BSs. The proposed scheme also gets rid of the burden of having to re-learn in the event that ground users move. The model's practicality is revealed by the fact that it takes average channel power gains as input rather than user locations. The outcomes demonstrated strong performance in comparison to other optimization methods. The study examines how to successfully use UAV BS to boost a cellular scheme's downlink network bandwidth. The best spot for flying BS on UAVs to help a macro-BS is chosen using an RL-aided technique. An approach is described to avoid accidents and restrict the motion of UAV-BSs into a specific area while several UAVs are performing exploration activities.

In [22], a single UAV is used as a relay node. Both an air-to-ground path loss model and a cellular-to-UAV route loss model have been applied to the backhaul and downstream connections, respectively. The basic goal is to place an aerial relay node in the best location possible in order to maximize the usage of resources by wireless devices. The overall amount of users throughput is optimized by using three different water filling and water modification methodologies to locate a comfortable position that maximizes throughput. In [26], experts use a multi-layer perceptron and long short-term memory approach to estimate the UAV positioning. In comparison to other systems, the effectiveness of the provided system is measured for user-throughput maximization and UAV position.

### *6.2.2 RIS-Assisted UAV Networks*

RISs regulate the wireless propagating environment by using a large number of programmable passive reflecting devices to achieve advantageous channel responses. For the multiple user system, improving the active beam-forming at the ground station as well as the passive beam-forming at RIS may considerably enhance the SINR efficacy of all network users, despite of if they are supported directly by the RIS or not. Furthermore, the suggested combined active and passive beam-forming configuration is also examined in various other software solutions, such as physical layer security (PHS), OFDM, NOMA, and simultaneous wireless information/power transfer (SWIPT). A common assumption in these and other prior RIS studies is that the perfect phase-shift model includes full signal reflection, which is impossible to achieve in practice because of hardware limitations. This assumption has been made in many other RIS studies as well. Passive reflecting elements have a non-uniform amplitude response due to their phase shift. Particularly at zero phase shift, amplitude tends to be lowest but monotonically rises and reaches unity magnitude at the phase shift of $\pi$ or $-\pi$, respectively. Due to the fact that an image current or currents from a virtual source are in phase with reflecting element currents as the phase shift nearing zero, an electric field and increased current flow are seen in each of the reflecting elements. This loss can't be prevented since it's mostly caused by the semiconductors, metals, and insulating substrates in RIS. This has been an issue for creating meta-surfaces based on reflection for a long time. However, this approach is unsuited for passively RIS and is also too expensive to execute on a practical level in [37].

Sixth-generation (6G) cellular communication and beyond fifth-generation (B5G) communication technologies such as RISs have huge capability to improve cellular communication [16]. Authors in [17], the RIS technology's key uses in wireless communication, as well as its key success factors over existing technologies, hardware architecture, and unique signal model are taken into consideration. The major difficulties in developing and deploying the latest RIS-supported Wi-Fi infrastructure that is hybrid are then discussed in contrast to previous existing infrastructure.

This work [20] demonstrates how a reconfigurable three-dimensional (3D) wireless relay system can benefit from the huge active and passive beam-forming gains offered by BS and RIS, in addition to the great movement of UAVs. The paired placement of UAVs, active beam formation and power allotment at BS, and passive beam forming at UAV-RISs are investigated as part of an optimization problem with the goal of maximizing the weighted sum rate. The combined optimization model is divided into a number of smaller tasks, each of which can be solved in order. The simulations demonstrate that the suggested solution performs significantly better than the benchmarks. RIS placed on UAV-assisted NOMA downlink systems is explored in [21].

To improve UAV effectiveness in Wi-Fi networks, RISs are utilized in addition to UAVs in [24]. Additionally, NOMA is applied to promote spectrum use while taking into account users that move irregularly. Optimizing UAV motions, UAV power allocation strategy, and RIS phase helps to address the challenge of minimizing energy usage. The method is provided as a declining deep Q network algorithm. However, the UAVs' optimized mobility has not been considered by the writers. In UAV-based OFDMA systems, RIS is sometimes utilized to boost system sum rate by utilizing both the significant beam-forming gain from the RIS and the UAV's high flexibility, according to [25]. In order to maximize the system total rate while taking into account the various QOS necessities for every user in a designed method, UAV trajectory, RIS scheduling, and resource allocation are merged into a non-convex optimization problem. So, to accelerate the planning process, upper and lower bounds for the described problem were provided using a parameterized estimator. An alternative optimal solution is used to address the lower bound optimization problem, and a comparison is made to the benchmark results attained by handling the upper bound issue.

The UAV acts as a relay for the work done by the authors of [27], who put a UAV communication system with RIS on a building to reflect signals generated by the station. In this setup, the decrypted signals are broadcast to the target by the UAV. In order to maximize the average achievable rate due to a shared UAV path and RIS's passive beam-forming structure, a new RIS-aided UAV network infrastructure is examined in [28]. The given non-convex task has two subtasks that need to be solved: passive beam formation and route optimization. Firstly, a closed-form phase-shift solution is designed for each specified UAV trajectory in order to coordinate the incoming traffic via various transmission links in phase. The best phase-shift solution is then obtained using the Successive Convex Approximation (SCA) approach with a suboptimal trajectory solution. The average attainable rate of the system is greatly increased by the suggested strategy, as demonstrated mathematically.

In dense urban environments, UAVs and RIS are being studied to determine if they can provide ground users with energy-efficient communication. The authors of [29] recommend an optimization scenario that ensures a specified QoS for the area's ground users in order to reduce energy consumption. Due to the complex nature of the optimization problem, a combined UAV path and RIS phase option to reduce transmit power produces excellent results with reduced complexity. Thus, the method described uses a successive convex approximation (SCA) to iteratively discover a joint optimal solution. In [30], the authors explored how changing the phase shifts of the RIS and transmission power can help the total rate in an multi-user MISO system network.

In spite of their promise, RIS-assisted UAV communication systems are constrained by factors including power consumption [31]. Due to its wireless function, the UAV is unlikely to be able to refuel itself, unlike conventional gadgets. As a result, the UAV consumes extra power, which is frequently in orders of magnitude more than the power required to communicate with it. As a result, efforts have been made to increase the UAV communication systems' efficiency in terms of energy consumption.

With a specified data restriction in mind, [32] optimized the UAV path, speed, and communication time in order to reduce energy usage. A number of efforts have been made recently to integrate RISs with UAV communication systems. According to [33], using a terrestrial RIS may help reduce UAV energy usage. However, [33] adopted the LOS model and used just one RIS to complete the assignment. Also, [34] optimized the RIS phase shifts, transmission, and UAV path to maximize the average reception power under a certain time limit. There was some discussion of numerous RISs in [34], but the trade-off between reception power and UAV flight power was not revealed. For example, the UAV may use extra flying power to optimize reception power. A multi-terrain RIS situation, in turn, encourages us to look for ways to reduce energy usage for UAVs.

The overall transmit power was reduced in work in [35], for example, by optimizing both the AP's active beam-forming technique as well as the RIS's passive beam-forming technique. The work in [36] describes how the system's passive beamforming efficiency is improved utilizing convergence point iterations and multivariate optimization approaches in an RIS-assisted MISO communication system, resulting in the highest possible transmission rate. It was suggested alternating optimization techniques for frequency-flat and frequency-selective channels to optimize the bandwidth utilization of an RIS-aided MIMO system. Weighted sum rate maximization under poor channel state information was studied in [36] using the stochastic sequential convex approximation (SCA) technique to improve active and passive beam-forming.

The attainable rate of deployed RISs with passively reflecting beam-forming using statistic channel state information (CSI) was studied in [38]. Coordinated transmitting and reflective beam-forming for optimizing the least averaged received signal-to-interference-plus-noise ratio (SINR) at users with transmit power limits was conceived and solved.

The authors in [39] optimized the RIS phase shifter coefficients and transmit precoding vector at the signal's sending end in order to increase the user's received signal strength. Numerous RISs were merged in a device-to-device communication architecture so that this communication network may be as energy efficient as possible. For multi-user clusters, the authors in [56] examined numerous UAV communications systems carrying RISs and grouped ground users. UAVs and RIS may be used to enhance the security of a communication network, according to the authors of those studies. Multi-user models were taken into account.

## 6.3 Proposed Resource Allocation Algorithm

### *6.3.1 System Model and Problem Formulation*

We consider a system model for maximization of system's energy efficiency by maximizing bits per joules of energy and our model consists of a UAV, an RIS and ground users (k). We have three transmission paths, GAG(t) is the direct transmission path from UAV to ground user (k), GAR (t) is the path from UAV to RIS, and GRG(t) is the path after reflection from RIS to ground user (k). Ground users are served either by direct transmission which is mode 1 (m = 1) or by the use of RIS which is mode 2 (m = 2). LOS shows line-of-sight communication between base station and user and N-LOS shows non-line-of-sight communication due to invisibility created between base station and user. xk,m(t) is the mode selection when user k selects mode m at time slot t. Rk,m(t) is data rate of ground user k which is achieved at time slot t with mode m. Pk,m(t) is the power allocated to each user k at time slot t with mode m. C1, C2, C3, and C4 are constraints of our system model which has certain limitations.

$$\max_{X,P} \sum_{t=1}^{\tau} \sum_{k=1}^{K} \sum_{m=1}^{2} \frac{x_{k,m}(t) R_{k,m}(t)}{P_{k,m}(t)} \tag{6.1}$$

$$C1 : \sum_{k=1}^{K} \sum_{m=1}^{2} X_{k,m}(t) \leq 1, \forall(t) \tag{6.2}$$

$$C2 : \sum_{k=1}^{K} \sum_{m=1}^{2} X_{k,m}(t) P_{k,m}(t) \leq P_{UAV}^{max}, \forall t \tag{6.3}$$

$$C3 : \sum_{t=1}^{\tau} \sum_{m=1}^{2} X_{k,m}(t) R_{k,m}(t) \geq R_{min}, \forall k \tag{6.4}$$

$$C4 : X_{k,m}(t) \in 0, 1, \forall k, \forall m, \forall t \tag{6.5}$$

where $X = \{x_{k,m}(t), \forall k, \forall m, \forall t\}$ and $P = \{p_{k,m}(t), \forall k, \forall m, \forall t\}$. The data rate $R_{k,m}$(t) for all users at each mode and at each time slot can be found as

$$R_{k,m}(t) = \log_2 \left(1 + \frac{P_{k,m}(t)|h_{k,m}(t)|^2}{\delta^2}\right) \tag{6.6}$$

where $\delta$ is the noise power and $h_{k,m}(t)$ is the channel and for each mode its expression is different as given below:

For mode 1:

$$h_{k,1}(t) = \frac{\sqrt{M_A}\beta_o}{d_k^{UAV-G}} \tag{6.7}$$

For mode 2:

$$h_{k,1}(t) = \frac{\sqrt{M_A} M_R \beta_o}{d_k^{UAV-G} d_k^{UAV-RIS}(t)} \tag{6.8}$$

### 6.3.2 Mode Selection sub-problem

Problem (6.1)–(6.5) is non-convex, non-linear fractional programming problem so we divide the whole problem into two sub-problems and solve these problem by using Dinkelbach algorithm. In sub-problem 1 we will find all the communication modes for given power allocation. The communication mode is for each user at each time slot (t). This sub-problem is subjected to constraints C1 and C4 and is linear binary integer programming problem and is written as

$$\max_X \sum_{t=1}^{\tau} \sum_{k=1}^{K} \sum_{m=1}^{2} \frac{x_{k,m}(t) R_{k,m}(t)}{P_{k,m}(t)} \tag{6.9}$$

C1 shows that each mode value for each user at each time slot is less than or equal to 1 and C4 shows that each mode value is either 0 or 1.

### 6.3.3 Power Allocation Sub-problem

In this sub-problem, we will find the power allocation for given mode of communication. Once the modes for each user at each time slots are selected then we will find the allocated power for that mode. This sub-problem is subjected to constraint C2 and C3 and is written as

$$\max_{P} \sum_{t=1}^{\tau} \sum_{k=1}^{K} \sum_{m=1}^{2} \frac{x_{k,m}(t) R_{k,m}(t)}{P_{k,m}(t)} \tag{6.10}$$

C2 shows that the allocated power of each mode is less than or equal to the maximum transmission power $P^{max}$ of base station which is UAV in our case. Similarly C3 shows that the data rate at each mode is greater than or equal to the minimum data rate limit $R_{min}$.

### 6.3.4 Proposed Algorithm

To achieve optimal solution of the formulated non-convex fractional programming problem, we divide the whole problem into two sub-problems and solve this problem by using Dinkelbach algorithm. In first sub-problem (Eq. 6.9) we find the communication modes for each user at each time slot and in the second sub-problem (Eq. 6.10), we find the allocated power at each mode. In our proposed approach, $\lambda$ and is represented as

$$\lambda = \sum_{t=1}^{\tau} \sum_{k=1}^{K} \sum_{m=1}^{2} \frac{x_{k,m}(t) R_{k,m}(t)}{P_{k,m}(t)} \tag{6.11}$$

By using Dinkelbach algorithm this equation is written in more simplified form as

$$\max_{P,\lambda} \sum_{t=1}^{\tau} \sum_{k=1}^{K} \sum_{m=1}^{2} x_{k,m}(t) R_{k,m}(t) - \lambda P_{k,m}(t) \tag{6.12}$$

The value of each mode with its data rate and power is stored for first iteration in the form of $\lambda$, i.e., Eq. (6.11). The value of x, R, and P in Eqs. (6.11) and (6.12) are same for same iteration. Once these values are put in Eq. (6.12) the algorithm starts and found some optimal values for given iteration. Similarly for next iteration $\lambda$ is updated by new values of x, R, and P in Eq. (6.11) and put values in Eq. (6.12), and so on. The algorithm takes long time to converge because the optimization problem is naturally convex. Algorithm 1 is summarized the Dinkelbach algorithm. The algorithm stops after a predefined termination criteria.

**Data**: Network parameters
**Result**: Optimal power allocation and mode selection
initialization;
Set $\lambda = 0$;
**while** *until termination condition* **do**
  Find optimal mode selection by using (9);
  Find optimal power allocation on these modes by using (10);
  Then find optimal data rates for mode and power allocation by using (6);
  Put above values in (11) to find value of $\lambda$;
  Using this $\lambda$, find the objective function value by using (12);
  Update $\lambda$;
**end**

**Algorithm 1:** Optimal Mode Selection and Power Allocation

## 6.4 Performance Evaluation

In this section, the performance of the proposed algorithm is evaluated. The simulation parameters are summarized in Table 6.1.

The results which are achieved from our algorithm are discussed one by one. Our main goal is to maximize the system energy efficiency by maximizing the bits per joules with the help of RIS (Fig. 6.3).

In Fig. 6.4, energy efficiency (bits per Joules) of our system is plotted for each time slot (t). Both graphs show different energy efficiencies (bits per Joules) values

**Table 6.1** Parameters and values

| Parameters | Values |
|---|---|
| Pmax | 40 dBm |
| Rmin | 50 bits/s |
| Users (k) | 5 |
| Time Slots (t) | 50 |
| Modes (m) | 2 |
| Noise | $10^{-15}$ |
| UAV antennas (MA) | 5 |
| RIS elements (MR) | 50 |
| Channel power gain (Bo) | −50 dB |

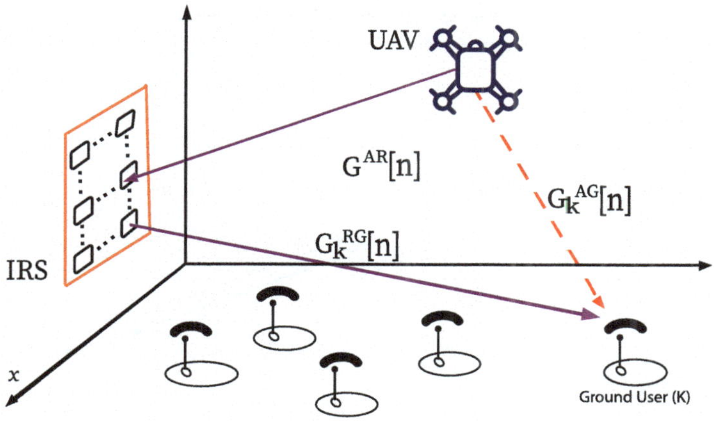


Fig. 6.3 System model

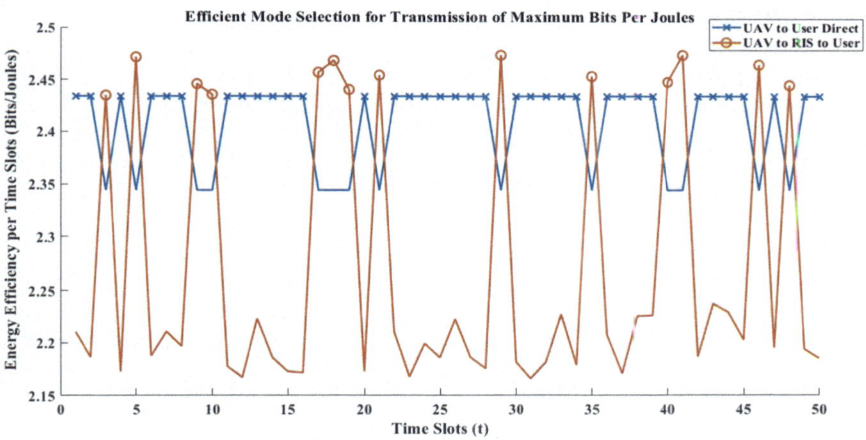


Fig. 6.4 Efficient mode selection for transmission of maximum bits per joules

in both modes (without RIS and with RIS). For LOS communication, bits are directly transmitted to user (mode 1) with some power. When there is no line-of-sight communication (mode 2) then more power is required to transfer data bits which results in decreasing bits per joule of energy so in this case we will use RIS which sends the intended data signal to user without losing power so in this way we will maximize bits per joule of energy by selecting the RIS path (mode 2). The modes are selected efficiently and only that mode is selected in each time slot (t) where maximum bits are transferred per joules of energy.

In Fig. 6.5, energy efficiency of our system is plotted against UAVs maximum transmission power (Pmax) in both modes (with RIS and without RIS). By our proposed approach, the energy efficiency is decreased by increasing value of Pmax

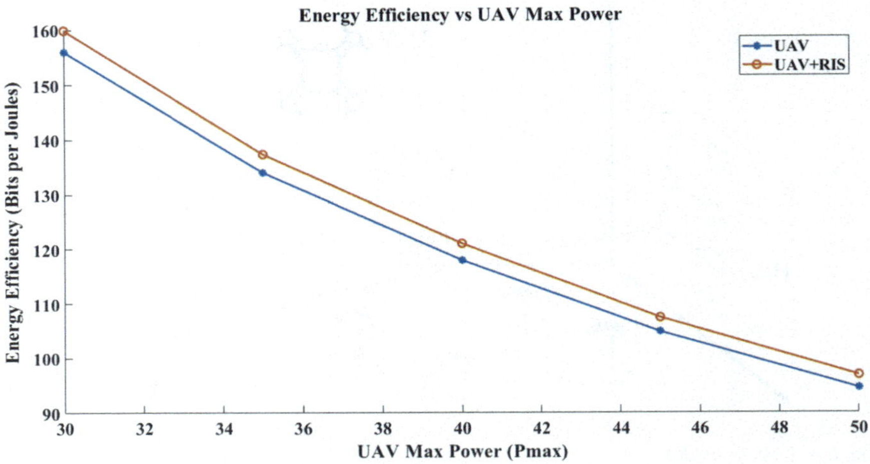


**Fig. 6.5** Energy efficiency versus $P_{max}$

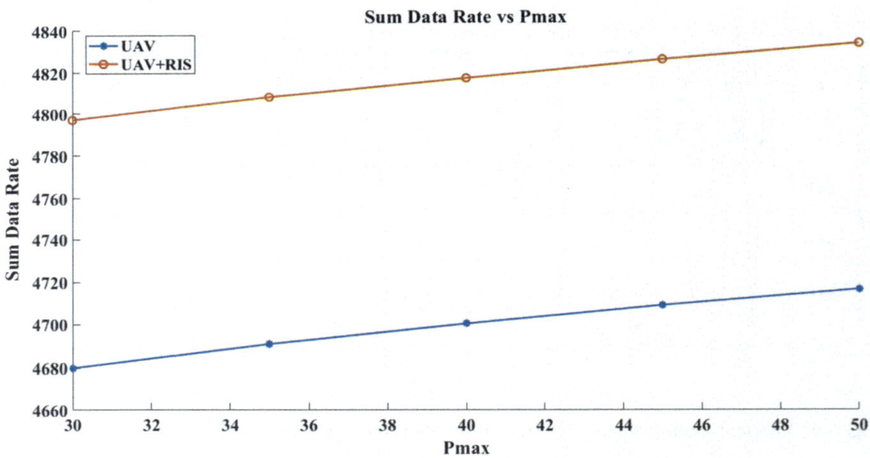


**Fig. 6.6** Sum data rate versus $P_{max}$

because by increasing Pmax, our objective energy-efficiency objective decreases. In both cases, we will get different values of energy efficiency by varying Pmax but better results are achieved by using RIS (mode 2) assisted with UAV. So in this way we can improve our system's efficiency by using RIS in our system.

In Fig. 6.6, sum data rates are plotted against maximum transmission power (Pmax). By increasing value of Pmax, the data rates are increasing exponentially in both modes (with RIS and without RIS) but better results are achieved by using RIS in our system. So in this way, with the help of RIS assisted with UAV we can improve our system's data rates by varying maximum transmission power (Pmax).

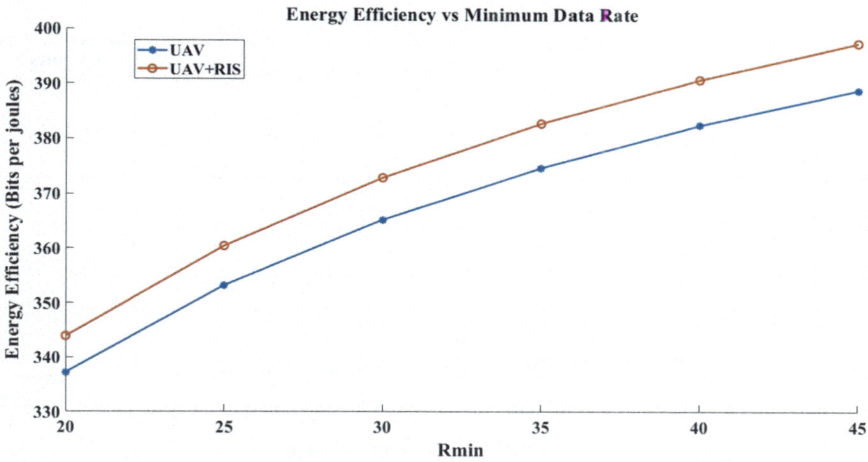


**Fig. 6.7** Energy efficiency versus $R_{min}$

In Fig. 6.7, energy efficiency is plotted against minimum data rates (Rmin). In this graph, energy efficiency is increasing by increasing minimum required data rate (Rmin) for our system. When Rmin is increased it means each user required more data rate than minimum data rate limit (Rmin) it means in the ratio R/P which is energy efficiency, the value of R is increasing but P remains same so R/P increases by increasing Rmin. In both cases, efficiency is increased but better results are achieved by using RIS assisted with UAV in our system as shown in graph. In Fig. 6.8, sum data rate is plotted against minimum data rates (Rmin). The plot depicts two schemes UAV and UAV+RIS, with the later showing better performance by yielding high data dates. For the analysis, the minimum data rate (Rmin) requirement is increased from 20 to 45 bps/Hz and for both the schemes the sum rate showed a decreasing trend. This trend follows the channel capacity limitations. As the minimum data requirement increases, the available channel capacity may not be sufficient to meet the demand, leading to a decrease in the sum rate of the system.

## 6.5 Conclusion

Due to flexibility and cost-effectiveness and high mobility, UAVs have recently gained popularity. It enhances coverage, reliability, and energy efficiency when utilized with a terrestrial base station. Flying base stations may be used in cellular networks. Reconfigurable intelligent surfaces (RIS) are the most cost-effective solution for networks beyond 5G and 6G due to their cheap cost and passive nature. They may be mounted on any building, rooftop, or UAV. The RIS-aided UAV networking design might improve wireless communications and handle the rising complexity of

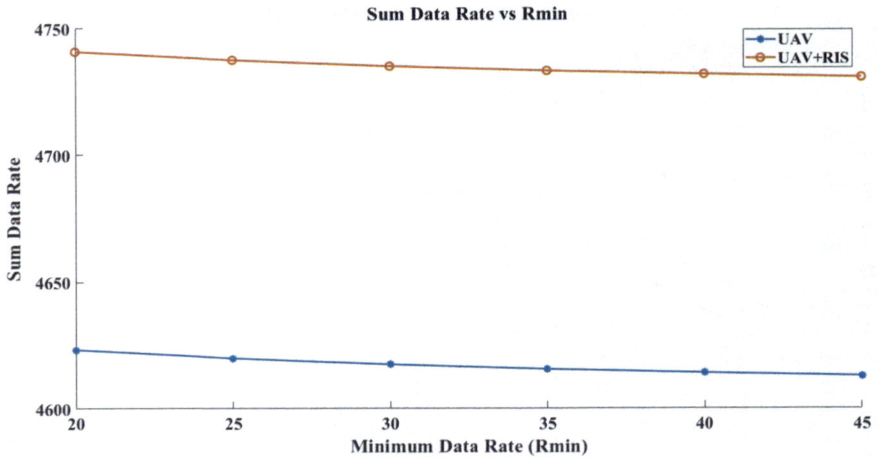


**Fig. 6.8** Sum data rate versus $R_{min}$

the wireless channel. When there is no LOS communication between base station and user, more power is needed to transmit data, we can use RIS which reflected the desired data signal from base station and sends it to the required user by adjusting its direction and phase. In our system, we used efficient mode selection with RIS-assisted UAV communication in which that mode is selected efficiently which has the maximum bits per joule value to make our system energy efficient.

**Funding** This work was supported by the Deanship of Scientific Research, Vice Presidency for Graduate Studies and Scientific Research, King Faisal University, Saudi Arabia [Grant No. KFU253968].

## References

1. Zeng Y, Zhang R, Lim TJ (2016) Wireless communications with unmanned aerial vehicles: opportunities and challenges. IEEE Commun Mag 54(5):36–42
2. Tugrul Cicek C, Gultekin H, Tavli B, Yanikomeroglu H (2018) UAV base station location optimization for next generation wireless networks: overview and future research directions. arXiv-1812
3. Hashida H, Kawamoto Y, Kato N (2020) Intelligent reflecting surface placement optimization in air-ground communication networks toward 6G. IEEE Wireless Commun 27(6):146–151
4. Nguyen KK, Khosravirad SR, Da Costa DB, Nguyen LD, Duong TQ (2021) Reconfigurable intelligent surface-assisted multi-UAV networks: efficient resource allocation with deep reinforcement learning. IEEE J Select Top Signal Proc 16(3):358–368
5. You C, Kang Z, Zeng Y, Zhang R (2021) Enabling smart reflection in integrated air-ground wireless network: IRS meets UAV. IEEE Wireless Commun 28(6):138–144

6. Alzenad M, El-Keyi A, Yanikomeroglu H (2017) 3-D placement of an unmanned aerial vehicle base station for maximum coverage of users with different QoS requirements. IEEE Wireless Commun Lett 7(1):38–41
7. Pogaku AC, Do D-T, Lee BM, Nguyen ND (2022) UAV-assisted RIS for future wireless communications: a survey on optimization and performance analysis. IEEE Access 10:16320–16336
8. Mohsan SAH, Khan MA, Noor F, Ullah I, Alsharif MH (2022) Towards the unmanned aerial vehicles (UAVs): a comprehensive review. Drones 6(6):147
9. Qin Z, Zhang X, Zhang X, Bingxian L, Liu Z, Guo L (2022) The UAV trajectory optimization for data collection from time-constrained IoT devices: a hierarchical deep q-network approach. Appl Sci 12(5):2546
10. Wu Q, Zhang S, Zheng B, You C, Zhang R (2021) Intelligent reflecting surface-aided wireless communications: a tutorial. IEEE Trans Commun 69(5):3313–3351
11. Cicek CT, Gultekin H, Tavli B, Yanikomeroglu H (2019) UAV base station location optimization for next generation wireless networks: overview and future research directions. In: 2019 1st international conference on unmanned vehicle systems-oman (UVS), pp 1–6. IEEE
12. Alzenad M, El-Keyi A, Lagum F, Yanikomeroglu H (2017) 3-D placement of an unmanned aerial vehicle base station (UAV-BS) for energy-efficient maximal coverage. IEEE Wireless Commun Lett 6(4):434–437
13. Shakhatreh H, Khreishah A, Alsarhan A, Khalil I, Sawalmeh A, Othman NS (2017) Efficient 3D placement of a UAV using particle swarm optimization. In: 2017 8th international conference on information and communication systems (ICICS), pp 258-263. IEEE
14. Zhang L, Fan Q, Ansari N (2018) 3-D drone-base-station placement with in-band full-duplex communications. IEEE Commun Lett 22(9):1902–1905
15. Zhang L, Ansari N (2018) On the number and 3-D placement of in-band full-duplex enabled drone-mounted base-stations. IEEE Wireless Commun Lett 8(1):221–224
16. Hashida H, Kawamoto Y, Kato N (2020) Intelligent reflecting surface placement optimization in air-ground communication networks toward 6G. IEEE Wireless Commun 27(6):146–151
17. Wu Q, Zhang R (2019) Towards smart and reconfigurable environment: intelligent reflecting surface aided wireless network. IEEE Commun Mag 58(1):106–112
18. Shakoor S, Kaleem Z, Do D-T, Dobre OA, Jamalipour A (2020) Joint optimization of UAV 3-D placement and path-loss factor for energy-efficient maximal coverage. IEEE Int Things J 8(12):9776–9786
19. Arani AH, Azari MM, Melek W, Safavi-Naeini S (2020) Learning in the sky: towards efficient 3D placement of UAVs. In: 2020 IEEE 31st annual international symposium on personal, indoor and mobile radio communications, pp 1-7. IEEE
20. Valiulahi I, Masouros C (2020) Multi-UAV deployment for throughput maximization in the presence of co-channel interference. IEEE Int Things J 8(5):3605–3618
21. Ali MA, Jamalipour A (2020) UAV placement and power allocation in uplink and downlink operations of cellular network. IEEE Trans Commun 68(7):4383–4393
22. El Hammouti H, Benjillali M, Shihada B, Alouini M-S (2019) Learn-as-you-fly: a distributed algorithm for joint 3D placement and user association in multi-UAVs networks. IEEE Trans Wireless Commun 18(12):5831–5844
23. Al-Ahmed SA, Shakir MZ, Zaidi SAR (2020) Optimal 3D UAV base station placement by considering autonomous coverage hole detection, wireless backhaul and user demand. J Commun Netw 22(6):467–475
24. Zhang L, Fan Q, Ansari N (2018) 3-D drone-base-station placement with in-band full-duplex communications. IEEE Commun Lett 22(9):1902–1905
25. Zhang L, Ansari N (2018) On the number and 3-D placement of in-band full-duplex enabled drone-mounted base-stations. IEEE Wireless Commun Lett 8(1):221–224
26. Wei Z, Cai Y, Sun Z, Ng DWK, Yuan J, Zhou M, Sun L (2020) Sum-rate maximization for IRS-assisted UAV OFDMA communication systems. IEEE Trans Wireless Commun 20(4):2530–2550
27. Munaye YY, Lin H-P, Adege AB, Tarekegn GB (2019) UAV positioning for throughput maximization using deep learning approaches. Sensors 19(12):2775

28. Li S, Duo B, Yuan X, Liang Y-C, Di Renzo M (2020) Reconfigurable intelligent surface assisted UAV communication: joint trajectory design and passive beamforming. IEEE Wireless Commun Lett 9(5):716–720
29. Deshpande AA, Vaca-Rubio CJ, Mohebi S, Salami D, De Carvalho E, Popovski P, Sigg S, Zorzi M, Zanella A (2022) Energy-Efficient Design for RIS-assisted UAV communications in beyond-5G Networks. In: 2022 20th mediterranean communication and computer networking conference (MedComNet), pp 158–165. IEEE
30. Nguyen BC, Hoang TM, Kim T (2021) On performance of two-way full-duplex communication system with reconfigurable intelligent surface. IEEE Access 9:81274–81285
31. Cho H, Choi J (2021) IRS-aided energy efficient UAV communication. arXiv:2108.02406
32. Zeng Y, Jie X, Zhang R (2019) Energy minimization for wireless communication with rotary-wing UAV. IEEE Trans Wireless Commun 18(4):2329–2345
33. Cai Y, Wei Z, Hu S, Ng DWK, Yuan J (2020) Resource allocation for power-efficient IRS-assisted UAV communications. In: 2020 IEEE international conference on communications workshops (ICC Workshops), pp 1–7. IEEE
34. Al-Hourani A, Kandeepan S, Lardner S (2014) Optimal LAP altitude for maximum coverage. IEEE Wireless Commun Lett 3(6):569–572
35. Wu Q, Zhang R (2019) Intelligent reflecting surface enhanced wireless network via joint active and passive beamforming. IEEE Trans Wireless Commun 18(11):5394–5409
36. Yu X, Xu D, Schober R (2019) MISO wireless communication systems via intelligent reflecting surfaces. In: 2019 IEEE/CIC international conference on communications in China (ICCC), pp 735–740. IEEE
37. Diamanti M, Tsampazi M, Tsiropoulou EE, Papavassiliou S (2021) Energy efficient multi-user communications aided by reconfigurable intelligent surfaces and UAVs. In: 2021 IEEE international conference on smart computing (SMARTCOMP), pp 371–376. IEEE
38. Zargari S, Farahmand S, Abolhassani B (2021) Joint design of transmit beamforming, IRS platform, and power splitting SWIPT receivers for downlink cellular multiuser MISO. Phys Commun 48:101413
39. Xie H, Jie X, Liu Y-F (2020) Max-min fairness in IRS-aided multi-cell MISO systems with joint transmit and reflective beamforming. IEEE Trans Wireless Commun 20(2):1379–1393

# Chapter 7
# Enhancing 6G Connectivity with Adaptive Deployment of Reconfigurable Intelligent Surfaces for Tethered and Untethered UAVs

**Mohamed Y. Selim, Shakil Ahmed, Ahmed Nazar, and Ahmed E. Kamal**

**Keywords** Unmanned aerial vehicle · Tethered and untethered · Reconfigurable intelligent surface

## 7.1 Introduction

In the realm of modern technology, the synergy between two cutting-edge domains, Reconfigurable Intelligent Surfaces (RIS), and Unmanned Aerial Vehicles (UAVs), is poised to redefine the landscape of wireless communication and remote sensing systems [1]. RIS technology, which involves the deployment of smart surfaces capable of manipulating electromagnetic waves to enhance communication and signal propagation, is on the cusp of a transformative integration with UAVs. This chapter serves as a technical exploration into the integration of RIS with both tethered and untethered UAVs, unlocking new frontiers in wireless connectivity, data transmission, and network optimization.

The fusion of RIS and UAVs represents a convergence of engineering excellence, where the precision of electromagnetic wave manipulation meets the versatility of autonomous aerial platforms. The possibilities of this integration extend far beyond

M. Y. Selim (✉) · S. Ahmed · A. Nazar · A. E. Kamal
Department of Electrical and Computer Engineering, Iowa State University, Ames, IA, USA
e-mail: myoussef@iastate.edu

S. Ahmed
e-mail: shakil@iastate.edu

A. Nazar
e-mail: amnazar@iastate.edu

A. E. Kamal
e-mail: kamal@iastate.edu

© The Author(s), under exclusive license to Springer Nature Singapore Pte Ltd. 2026 
I. Ahmad et al. (eds.), *Reconfigurable Intelligent Surfaces for Unmanned Aerial Vehicle Communication*, Unmanned System Technologies,
https://doi.org/10.1007/978-981-96-9662-8_7

the theoretical realm, offering tangible solutions to critical challenges in wireless communication, remote sensing, and data delivery [2].

At its core, RIS is a metamaterial-based technology. These intelligent surfaces are capable of dynamically altering their electromagnetic properties to control the reflection, refraction, and absorption of radio waves across various frequencies. Such precise manipulation of electromagnetic fields promises unparalleled gains in signal quality, coverage, and spectral efficiency [3].

UAVs, on the other hand, are evolving from mere surveillance tools into sophisticated airborne nodes within wireless communication networks. Their mobility, versatility, and ability to operate in diverse environments make them ideal platforms for deploying RIS technology in real-world scenarios. As UAVs equipped with RIS soar through the skies, they become catalysts for next-generation wireless systems, enabling applications ranging from enhanced rural broadband access to tactical battlefield communications.

Tethered UAVs (T-UAVs), or UAVs connected to the ground/base station via a cable, offer unique advantages in terms of prolonged flight times and stability. By incorporating RIS technology into T-UAV systems, i.e., T-UAV$^{RIS}$, these aircraft can be deployed in various scenarios, such as disaster response and surveillance, with enhanced communication capabilities. The RIS-equipped tethered UAVs can dynamically adjust their communication patterns to establish more reliable and efficient connections with other network nodes or ground stations.

Conversely, Untethered UAVs (U-UAVs) can benefit from RIS integration by improving their wireless communication and data transmission/reflection capabilities. These UAVs can leverage RIS technology to enhance signal strength and reduce signal interference. The integration of RIS with U-UAVs, i.e., U-UAV$^{RIS}$, opens up possibilities for establishing ad hoc networks in remote or challenging terrains where traditional communication infrastructure may be lacking.

The integration of RIS and UAVs has the potential to offer a number of unique advantages, including

- Enhanced coverage and capacity: RIS can be used to extend the coverage of wireless networks and increase their capacity, especially in challenging environments such as urban areas and rural areas with poor infrastructure. UAVs can be deployed to dynamically adjust the placement of RIS elements, further improving the performance of the network.
- Reduced interference: RIS can be used to reduce interference between different users and devices in a wireless network. This is particularly important in dense networks, where interference can be a major problem. UAVs can be used to deploy RIS elements in strategic locations to minimize interference.
- Improved security: RIS can be used to improve the security of wireless networks by making it more difficult for eavesdroppers to intercept signals. UAVs can be used to deploy RIS elements in dynamic locations, further enhancing the security of the network.

The emergence of UAVs assisted by RISs expands solutions for networking and communication problems in next-generational networks. The signal-manipulative abilities of the RIS complement the high degrees of freedom of UAVs. The UAV can use intelligent trajectory planning and positioning to create Line-of-Sight (LoS) links in Non-LoS (NLoS) areas by expanding the reflective capabilities of RISs mounted in a strategic and optimized location. Communication links with efficient resource allocation improvements as positioning and phase change to an optimized location for various approaches such as enhanced spectrum sharing, massive access, and increased channel capacity [4].

RISs have shown immense potential in enhancing wireless communications systems. However, traditional RIS implementations relying solely on reflection-based techniques have posed limitations, including signal loss and reduced flexibility. To overcome these challenges and comply with 3GPP standards, a new approach that integrates active and passive elements within RIS is proposed. Meanwhile, studies of actively simultaneously transmitting and reflecting RISs (STAR-RISs) with UAVs to address the evolving challenges in wireless communication networks are currently being addressed by the research community [5, 6].

The advantages of UAV and RIS integrations bring us to the motivation for our chapter, where our primary focus is on the integrations of RIS with UAVs and the variations of both technologies. Our chapter introduces the integrations of RIS technologies with tethered and untethered UAV technologies. This chapter provides information regarding existing solutions that utilize the integrations of the two technologies and their deployment schemes and shows that this has strong potential benefits in enhancing communications for the future of 6G connectivity. This chapter also provides some state-of-the-art work that achieved compelling results.

In our exploration of the integration of Reconfigurable Intelligent Surfaces (RIS) with both tethered and untethered UAVs, we have delved into two compelling case studies. The first case study, focusing on U-UAVs equipped with RIS (U-UAV$^{RIS}$), offers an intriguing glimpse into the advantages of mounting RIS panels on these agile aerial platforms. This approach not only increases the probability of Line-of-Sight (LoS) connectivity but also empowers 360° full-angle reflection.

Our second case study, on the other hand, highlights the immense potential of RIS technology in addressing the challenges posed by signal blockages in wireless communications with the aid of U-UAVs. RIS technology significantly elevates the quality of communication links for U-UAVs, intelligently mitigating path loss, signal fading, and interference.

In this case study, a rate maximization problem is formulated subject to the UAV mobility constraint and the main aim of this problem is to find the U-UAV optimal trajectory for a given flight time.

These case studies exemplify the versatility and innovation brought about by the fusion of RIS technology with UAVs, promising a future where wireless communication and remote sensing reach new heights of efficiency and effectiveness.

## 7.2 Exploring Tethered and Untethered UAV Solutions

Our chapter covers two deployment schemes for UAVs: Tethered UAVs (T-UAVs) and Untethered UAVs (U-UAVs). The main difference between the two schemes is that T-UAVs employ a tethered connection to the base station (BS) for a constant power supply and direct fronthauling/backhauling (if needed), whereas U-UAVs have higher degrees of freedom with higher flexibility, a limited power supply and less stable data exchange with the BS. This chapter section explores the different mounting schemes and their advantages to provide insight into solutions for both schemes.

### *7.2.1 Untethered UAVs (U-UAVs)*

U-UAVs directly deliver communication information and data between senders and receivers. With strategic path planning and positioning, this technology supports and improves connection links between base stations, specific nodes, and LoS areas. U-UAVs' advantageous property is the dynamic movement and creation of coverage areas. The ability to fly autonomously or have manned flight that creates coverage areas in non-LOS areas makes U-UAVs an alluring deployment item for communication network applications. The use of U-UAVs has potential due to the increase in the likelihood of establishing LoS links and the strong backhaul and access links. In heavy congestion areas, the flexible nature of U-UAVs improves and enhances coverage and quality of service (QoS).

Different sizes of U-UAVs are used for different missions and deployment criteria; some use even small U-UAVs in swarms to improve data communication and create dynamic LoS nodes [7, 8]. U-UAV swarms and distributed networks promote the abilities of enhanced throughput, scalability, and increased efficiency due to multiple established LoS links [8]. The favorable aspect of the dynamic movement of U-UAVs supports the idea of swarms and allows for efficient resource allocation [8].

Resource allocation is a topic of interest for swarms and distributed joint U-UAV networks. Power control and hovering altitude can be jointly optimized in U-UAV networks while considering cross-tier interference from space-air-ground heterogeneous networks [9]. This joint and distributed approach allows U-UAVs to hover in locations with NLoS links to establish LoS links. Network throughput was greatly enhanced using joint and distributed two-stage tactics while keeping resource allocation concerns within expected boundaries [9].

### *7.2.2 Tethered UAVs (T-UAVs)*

Tethered platforms are attached to a ground station by a heavy-duty cable that gives the flying platform power and fronthauling/backhauling. Although it always sounds uncanny for a flying platform to be tethered, this has some advantages; the most relevant one is the stable power source, allowing unlimited flying time. T-UAVs have overall lower operating costs than U-UAVs; however, their main drawback is limited mobility [10].

Terrestrial and cellular satellite services are costly and require time for deployment. In contrast, airborne communication systems, like UAVs, have addressed these challenges and demonstrated promising results. Nevertheless, despite the potential of these free-flying platforms, they need help with endurance and data capacity. Conversely, T-UAVs surmount these limitations by ensuring a continuous supply of power and data through the tether. Additionally, they offer cost-effectiveness and environmentally friendly attributes. Tethered platforms have diverse applications, including energy generation, entertainment, scientific research, public safety, and disaster relief. It is worth noting that only rotary-wing UAVs can be tethered due to their hovering capability.

Tethered UAVs offer a unique set of technical advantages. T-UAVs are fast to deploy, with a deployment time of within a few minutes. The tether provides a direct power source, ensuring a constant and reliable energy supply, allowing for extended flight times without the need to return for recharging or battery swaps. This constant power supply is crucial in applications like persistent surveillance, where uninterrupted aerial coverage is vital. The tether is a secure data connection, enabling high-bandwidth data transfer, low-latency communication, and real-time telemetry, enhancing the UAV's operational capabilities. T-UAVs also benefit from increased stability due to the fixed anchor point, reducing motion and drift during hovering, which is essential for precise positioning and data collection.

### *7.2.3 Differences Between U-UAVs and T-UAVs*

The following subsection describes the main differences between U-UAVs and T-UAVs at a high level:

- **Flight Time:** T-UAVs have extended flight times and receive a consistent power supply while the BS is alive. U-UAVs are constrained by the onboard battery capacity, typically ranging from a few minutes to an hour, depending on the UAV's type, size, and payload [7, 11].
- **Deployment and Setup:** T-UAVs are connected to a ground station by a physical link [12]. The tether provides power supply and data communication between the UAV and the BS. U-UAVs operate wirelessly with respect to the BS. They are equipped with onboard batteries and use wireless communication [7].

**Table 7.1** Comparing untethered versus tethered UAVs

| Properties | T-UAVs | U-UAVs |
|---|---|---|
| X-Hauling | Wired/Wireless | Wireless |
| UAV used | Rotary wing | Rotary wing, fixed wing<br>Vertical Take-Off and Landing (VTOL) |
| Battery capacity | Consistent power supply | Limited |
| Ease of deployment | Complex | Fast |
| Deployment speed | Slow | Fast |
| Degrees of freedom | Limited | High |
| Latency | Low latency | Depends |
| Bandwidth | High | Low |
| Regulations to follow | Terrestrial and airspace | Airspace |

- **Payload:** T-UAVs can carry heavier and more sophisticated payloads due to the consistent physical link to the ground station, whereas U-UAVs are often limited by their size, weight, and battery capacity [7, 13, 14].
- **Degree of Freedom and Mobility:** T-UAVs have limited mobility as the length of the tether cable restricts them. They are well suited for stationary or semi-stationary applications [13]. U-UAVs offer high mobility and flexibility, allowing them to cover large areas and access hard-to-reach locations, making them suitable for various dynamic applications [7, 14, 15].

A summary of high-level differences between T-UAV and U-UAV technologies is seen in Table 7.1, where the differences are further supplemented by the respective subsections.

## 7.3 From Early to Modern RIS

The development of Reconfigurable Intelligent Surfaces (RIS) represents a paradigm shift in how we perceive and utilize technology. Over the years, RIS has emerged from its nascent stages to become a transformative force that promises to redefine wireless reflections. One of the main silver linings of an RIS is abiding by a UAV's Size, Weight, and Power (SWaP) constraints. An RIS device's low cost and design make it a clever constituent to use in the merger of $UAV^{RIS}$ communication and networking. The favorable physical attributes provide a path for various mounting techniques to further expand RIS advantages. The mounting techniques are

- **Mounting RIS on a UAV:** In some applications, such as in [16], an RIS is mounted on a UAV to adjust phase coefficients and reflect signals while mobile. The UAV's

**Fig. 7.1** Different mounting methods for RIS technologies

mobility and the RIS's configurability allow for optimal phase change within the dynamic positioning of the UAV.

- **Mounting RIS on Vehicles:** Another possibility for mounting is using manned vehicles to adjust and move the RIS around to establish links with LoS. An example is in [17], where Aerial Intelligent Reflecting Surface (AIRS) is implemented to improve communication. The idea is to achieve higher ground or improve signals in non-LoS areas and strengthen the channel compared to terrestrial mounting methods.
- **Mounting RIS Terrestrial Surfaces:** Some solutions mount RIS devices on terrestrial surfaces [18]. The terrestrial surfaces are chosen based on the locations of the surfaces that optimize signal power and SNR. Examples of such locations are building surfaces or specific ground locations.

Figure 7.1 displays a visual representation of the mounting methods associated with RIS approaches.

### *7.3.1 Passive RIS*

The passive RIS does not amplify an incoming reflected signal and modifies the phase shifts and reflection angles described above. Most available applications and research papers have been completed using passive RIS. Passive RIS modules are subject to the multiplicative fading effect, which implies that the path loss in the transmitter-to-receiver link is the product of the RIS-to-transmitter and RIS-to-receiver link [19]. The path losses would be equivalent to the sum but not the product. The capacity gains of the passive RIS in real-world scenarios are negligible; hence, the ability for active RIS to thrive is introduced [19, 20].

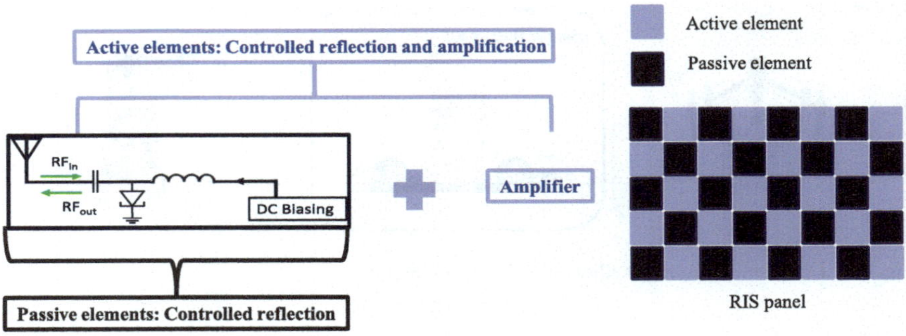


**Fig. 7.2** Active and passive elements

## *7.3.2 Active RIS*

As opposed to passive RISs that merely reflect signals, active RISs enhance the reflected signals by incorporating amplifiers into their elements. These amplifiers boost incoming signals but come at the cost of increased power consumption, making active RISs operate at a higher power level than passive RISs. An RIS can have a mix of active and passive elements, and their selection can be optimized based on minimizing the overall energy consumption of the RIS [19].

Integrating active elements into RIS introduces dynamic control over signal propagation. Active elements enable signal amplification, modulation, and adaptive beamforming, expanding the capabilities of RIS beyond passive reflection. Passive elements play a vital role in signal reflection and manipulation, contributing to the system's overall performance in compliance with 3GPP standards. Figure 7.3 shows the integration of active and passive elements into the same RIS panel (Fig. 7.2).

## *7.3.3 STAR-RIS*

Simultaneously Transmitting And Reflecting (STAR) RIS indicates that these RIS elements can reflect incoming signals and transmit their signals simultaneously, contributing to the overall communication process. This property means that STAR-RIS can actively reflect signals for improved reception and transmitting/refracting signals to enhance the transmission process, as illustrated in Fig. 7.3, effectively optimizing wireless communication. This capability can significantly benefit scenarios such as link establishment between base stations and users and potentially improve the efficiency and performance of wireless networks [5, 6].

**Fig. 7.3** STAR-RIS

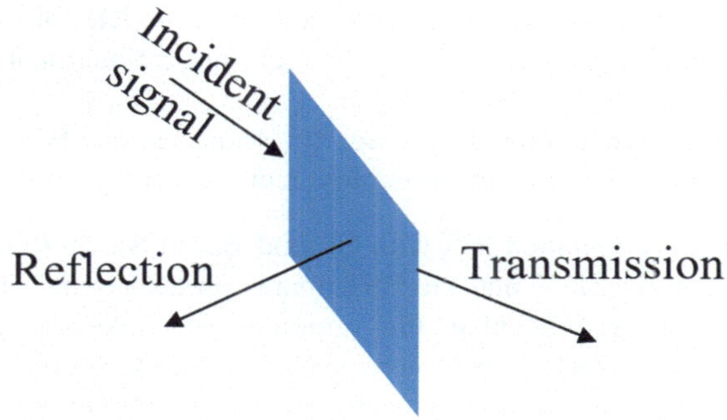


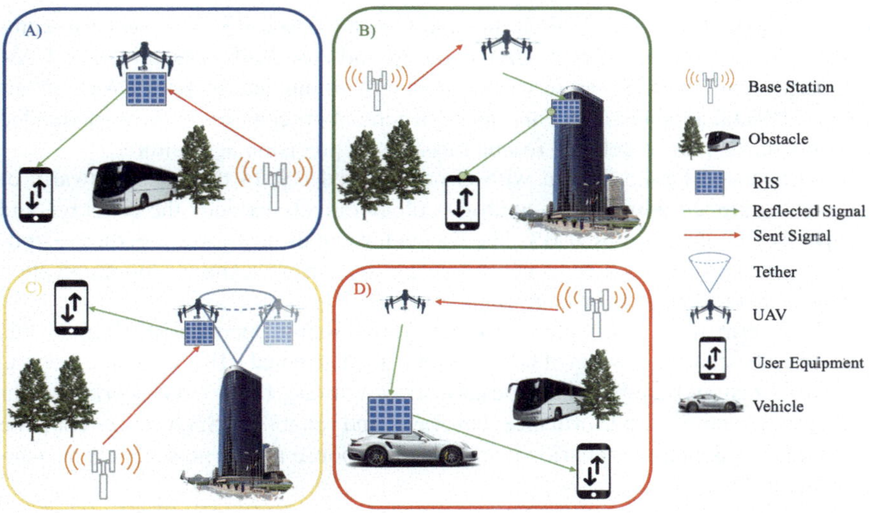


**Fig. 7.4** Different mounting methods for the integration of UAV$^{RIS}$. **A** shows mounting an RIS on a U-UAV, **B** shows terrestrial mounting of an RIS on a building facade, **C** shows an RIS mounted on a T-UAV, and **D** shows vehicular mounting of an RIS

## 7.4 The Integrations of UAV$^{RIS}$

There are multiple mounting methods for UAV$^{RIS}$ technologies. Some of the mounting methods include an RIS mounted directly on a UAV, RIS mounted on a building or terrestrial plane, and RIS mounted on a vehicle [21–23]. Figure 7.4 displays four examples of the U-UAV$^{RIS}$ and T-UAV$^{RIS}$ mounting methods for communication between UE and BS in the presence of obstacles to improve communication and data transmission within a network.

As for the reasoning and benefits of mounting RISs on UAVs, one of the primary motivations is the enhanced flexibility and adaptability this combination provides, which are critical for 6G-enabled communication systems. The UAV's mobility allows for dynamic repositioning of the RIS, aiding in overcoming obstacles or optimizing signal pathways in real time, a functionality that static RIS installations lack.

Different mounting cases for scenarios involving a UAV, an RIS, and a Base Station (BS) in various deployment scenarios are presented below:

- UAV-mounted RIS with Ground-Based BS: In this scenario, the RIS is mounted on the UAV, and the BS remains on the ground. The UAV with the RIS can be deployed for enhancing communication links in scenarios where the ground-based BS's coverage is limited, such as disaster recovery or surveillance missions. The UAV can adaptively position the RIS to optimize signal strength and quality for ground-based users.
- UAV-mounted RIS with UAV Relay and Ground-Based BS: This scenario combines the capabilities of both UAVs and ground-based infrastructure. The UAV, equipped with the RIS, can serve as a relay for communication between a ground-based BS and another UAV. This can be beneficial in scenarios requiring extended coverage, such as search and rescue missions or precision agriculture.
- RIS-Enhanced Drone Swarm with Ground-Based BS: In this case, a swarm of drones equipped with RIS technology collaboratively extends the coverage and capacity of a ground-based BS. The RIS-enhanced drones work together to optimize signal propagation, reduce interference, and ensure seamless communication in crowded urban areas or large-scale events.
- UAV Swarm with Distributed RIS and Autonomous Decision-Making: In this advanced scenario, a swarm of UAVs, each with distributed RIS panels, autonomously makes decisions based on real-time network conditions. These UAVs work together to optimize network performance, coverage, and reliability. Such setups could be suitable for dynamic scenarios like disaster response and large-scale event management.

These mounting cases illustrate the versatility and adaptability of RIS and UAV technology in different deployment scenarios, enabling enhanced wireless communication and coverage in a wide range of applications and environments. The choice of scenario depends on the specific requirements and challenges of each use case.

## 7.5 State of the Art: UAV$^{RIS}$

The following section in our book chapter provides the details of merging RIS and UAV technologies based on existing solutions from the realm of communications. The main goal of this section is to provide insight into the advancements made by using a UAV and an RIS together to push the technology to be a state-of-the-art solution scheme.

### 7.5.1 U-UAV$^{RIS}$ in Transmission Strategies

UAV trajectory is a crucial consideration for any UAV approach. Positioning and alignment make the difference between achieving optimal and subpar optimization goals. Leveraging instantaneous and statistical channel state information under Rician Fading conditions can improve overall system performance by employing a hybrid offline-online U-UAV$^{RIS}$ approach [24]. Offline conditions refer to the state where U-UAV ground-channel states are very random and dynamic, data loss rates are higher due to lack of adaptability, and the online conditions are the converse of the offline conditions. The maximum achievable rate is jointly optimized through the RIS phase shift and U-UAV trajectory based on the statistical channel state information for the offline phase. In the online phase, the U-UAV transmit beamforming and user equipment (UE) scheduling are altered based on the immediate channel state information. The optimization problem utilizes stochastic successive convex approximation techniques to provide a solution. Simulations and numerical results show that a hybrid approach utilizing U-UAV$^{RIS}$ and an online–offline approach based on variations of channel state information efficiently solves the maximal average achievable rate problem under Rician Fading conditions [24].

Transmission strategy, channel estimation, and data transmission are important aspects of optimal communication. One method of determining the optimal channel estimation, transmission strategy, and data transmission is through a frame-by-frame analysis of incoming data to an RIS. Multi-task learning can be used to approximate a solution to an optimization problem and is modeled as a mixed-integer non-linear program [25]. The program outputs the maximum overall throughput of the system. A terrestrial RIS assists the UAV in optimizing the data transmissions within this model. The channel estimation is processed at the UAV level, and the RIS optimizes the transmission throughput. This division of tasks allows for less processing power at the UAV level due to its low battery life and faster processing at the RIS since channel estimation is complicated. The allocation of RIS elements, phase shifts, and UAV position is optimized through the output of the multi-task learning model. By utilizing a learning model at the RIS level, communication is improved greatly according to the results in the paper and, therefore, achieves the goal of merging RIS and UAV, enhancing QoS.

Transmission strategies are pivotal in enhancing UAV$^{RIS}$ approaches. By intelligently coordinating the transmission of signals from UAVs to RIS and subsequently to their intended destinations, these strategies can optimize the efficiency of data delivery and energy consumption. Strategic signal beamforming, power allocation, and resource scheduling techniques can significantly enhance the overall performance of UAV$^{RIS}$ systems. These approaches improve data throughput, reduce interference, and maximize the utilization of RIS-enabled signal reflections, making UAV$^{RIS}$ communication solutions more effective and reliable for various applications.

### 7.5.2 U-UAV$^{RIS}$ and Positioning and Coverage

Integrating U-UAV$^{RIS}$ enhances positioning and coverage for improved communication. RIS technology optimizes signal reflection and redirection, allowing for precise control of signal beams. UAVs provide dynamic mobility and flexibility, enabling them to be deployed strategically to optimize coverage areas and adapt to changing communication needs. This combination of U-UAV$^{RIS}$ technologies results in more reliable and extensive network coverage, improving communication quality and enabling better positioning for various applications.

U-UAV$^{RIS}$ positioning and coverage optimization are important considerations for any U-UAV$^{RIS}$ approach. One goal of positioning and coverage optimization above is to maximize the data transfer rate by jointly optimizing the RIS phase shift and the 3D space of the U-UAV, where energy efficiency and positioning are the criteria to be enhanced. Deep Q-network and deep deterministic policy gradient algorithms are utilized through the use of deep reinforcement learning [26]. The machine learning algorithms result in a solution for the U-UAV trajectory and passive beamforming. The phase shift of the RIS is altered with respect to the outcome of the learning algorithms to optimize the alignment of the U-UAV to BS signal reflection angles. The proposed solution improved energy efficiency by aligning and positioning the U-UAV accordingly while satisfying the BS data transmission requirements.

U-UAV positioning and alignment depend entirely on the application to solve the problem. Some approaches optimize the positioning for a minimal bit-error rate, and others with respect to the optimal energy efficiency of a system [11]. Other approaches utilized assisted devices for path planning as this deployment aspect is crucial to securing the solution's effectiveness to an approach [11]. Details of such positioning can be seen in the UAV survey in [11].

Maintaining LoS links and increasing coverage areas is done by optimizing the deployment and positioning of a UAV with respect to a BS or RIS. A collection of U-UAV$^{RIS}$ is deployed within high-density areas to complete the task mentioned above [27]. Efficiency and performance are crucial for next-generational networks, specifically for the topic of interest, 5G and mmWave frequencies. A framework was developed for sustaining and maintaining LoS links in high-density areas using RIS-mounted U-UAVs [27]. A consensus-based approach increases the coverage area and QoS for all dynamic UE, while the RIS reflects signals to improve efficiency and communication within the high-density areas. Tests were conducted on the consensus algorithm-utilized effective navigation framework defining several clusters for simulation within the scope of a dynamically changing environment with LoS and NLoS areas [27]. The simulation results provide evidence that the consensus algorithm follows through with the goal of the approach.

### 7.5.3 U-UAV$^{RIS}$ in Energy Efficiency

Energy efficiency is enhanced by U-UAV$^{RIS}$ in a broad range of applications. This synergy optimizes the use of resources, minimizes energy consumption, and improves overall system performance. RIS technology facilitates better signal control and reflection, reducing power wastage, while UAVs provide flexibility in deployment and energy-efficient data collection. RIS and UAVs contribute to more sustainable and efficient wireless communication systems.

Increased energy efficiency is a favorable outcome of optimizing U-UAV$^{RIS}$ approaches in many ways. In mobile edge computing (MEC) applications, the quality of optimal data exchange and energy efficiency highly affects the outcomes of the approach, and as such, a U-UAV$^{RIS}$ can be utilized to increase the quality of the outcomes. Deploying a U-UAV$^{RIS}$ can aid the communication between ground user equipment and MEC servers where the trajectory and path planning of the U-UAV is optimized jointly with the RIS passive beamforming and MEC server resource allocation to improve energy efficiency of the data exchange [28]. The idea of energy efficiency is achieved by finding the ratio between offloaded tasks between the devices in the communication network and the weighted total energy of the system. Tasks are determined by bits offloading and completed per user. A high-performance suboptimal solution was achieved by leveraging the optimization problem into two subproblems and jointly optimizing the RIS phase shift matrix, the UAV's trajectory, and the number of bits in offloaded tasks where the solution showed considerable improvements in energy efficiency using a U-UAV$^{RIS}$ compared to other approaches that exclude the U-UAV$^{RIS}$[28].

Wireless power transfer by utilizing U-UAV$^{RIS}$ mentioned in [29] allows improvement in EE by optimizing the U-UAV trajectory and hovering location and the RIS reflection coefficient. A successive convex approximation of a minorization-maximization algorithm is proposed, which optimizes the minimum charged energy ground sensors in calculating reflection coefficients. With the optimization of the reflection coefficients comes the joint optimization of the U-UAV trajectory and hovering location. The U-UAV radiates radio-frequency signals only at several hovering locations optimized for minimizing U-UAV energy consumption. Integrating the RIS into wireless power transfer shows that the U-UAV energy consumption decreases, meaning that the overall energy requirements for the application are reduced considerably considering both U-UAV$^{RIS}$ [29].

### 7.5.4 U-UAV$^{STAR\text{-}RIS}$ Communications

Simultaneously, Transmitting and Receiving (STAR) RIS differs from typical RISs. STAR-RIS technologies transmit and receive signals on either side of the surface to achieve 360-degree wireless coverage [5]. STAR-RIS requires the design of complex signal transmission and receiving schemes to achieve desired results for enhanced communications [5].

STAR-RIS can assist U-UAVs to increase coverage of the U-UAV and attain enhanced network properties such as increased robustness and optimized trajectory design with more degrees of freedom. The desirable 360-degree wireless coverage from STAR-RIS allows for full utilization of the U-UAV and allows ground users to receive communications from the non-reflective side of a conventional RIS [30]. The 360-degree coverage is adapted for high-mobility U-UAV communications while coming into contact with dynamic and random obstacles existing in the environment. As such, machine learning solutions can be introduced to increase the robustness of the STAR-RIS-assisted UAV approaches and improve the learning of the dynamic network environment [30].

Fading can be a source of challenges for STAR-RIS, leading to delayed and poor channel state information acquisition. As such, utilizing STAR-RIS with a U-UAV can improve the sum rate of Non-Orthogonal Multiple Access (NOMA) UAV networks by jointly optimizing the U-UAV's location based on its hovering characteristic, with the transmit power and transmission/reflection beamforming without channel state information for U-UAV trajectory planning which is limited by the QoS requirements of the network [31]. The main optimization problem is broken down into two subproblems dealing with the UAV optimal location and the transmit power and beamforming of the RIS, which is then solved iteratively for an optimal solution to address the QoS requirements. This approach allows for a considerable improvement in sum rate for NOMA UAV networks without the need for channel state information using U-UAV$^{\text{STAR-RIS}}$.

While passive RIS primarily focuses on signal reflection, STAR-RIS can actively transmit and reflect signals. This dual capability empowers the UAV-RIS network to enhance signal quality and coverage through reflection and proactively transmit data to desired destinations. The active nature of STAR-RIS greatly enhances the network's flexibility, adaptability, and overall performance, making it a valuable tool for applications where both transmission and reflection are crucial for efficient and reliable communication.

### 7.5.5 T-UAV$^{RIS}$ in High-Throughput Communication

Using a T-UAV comes with many advantages. The two most notable advantages are the provision of a tether with a fiber link from the BS to the UAV and a constant power supply for continuous backhauling. The advantages of utilizing T-UAVs are the power consumption and life of the devices. The power consumption is heavily decreased since some processing power is handled at the linking station. The cost of this implementation is considerably less than a U-UAV due to the lack of reason for optimizing the location of the T-UAV midflight. This reason allows for deploying T-UAVs within smaller rural areas to increase and strengthen signal communication. The link from the station is not prone to interruption. The link allows for higher data traffic, which means a higher capacity through the tether link for larger amounts of payload data. T-UAVs also free up resource blocks usually reserved for U-UAV applications, which can be utilized in other network applications related to the environment [32].

In the case of T-UAV$^{RIS}$ for backhauling, if there is power within the system model of interest for a specific problem, then the concept of backhauling is negligible; as such, this emphasizes that RIS can be used to improve communication hand-in-hand with a T-UAV. One application that utilized the power of the tethered link is seen in CURE, which is a framework where radio-frequency energy harvesting (RFEH) is merged with a massive cell-free MIMO (CFmMIMO), T-UAV, and RIS network to enhance the wireless power transfer to Internet-of-Things (IoT) devices [33]. The problem addressed in this approach is enhancing the RFEH through an RIS-assisted network where the Cell-free access points are mounted on T-UAVs over the coverage area. The wireless power transfer from the T-UAV is then reflected off the RIS with the metasurface's manipulative nature to optimize the phase angle shift and maximize the signal transmitted to an IoT device. The approach was tested on a max-min fairness algorithm, which resulted in higher energy transfer and data transmission.

T-UAV$^{RIS}$ ensures continuous flight and power supply, allowing for extended network coverage and persistent data transmission through the tethered link. Simultaneously, the RIS technology optimizes signal reflection and redirection, enhancing data quality and reducing interference. This cooperative strategy extends the UAV's operational duration and optimizes signal propagation, resulting in a more reliable, high-throughput communication network that can efficiently meet the data demands of various applications.

### 7.5.6 T-UAV$^{RIS}$ in Energy Efficiency

Due to the consistent power advantage in T-UAV, energy efficiency can be enhanced following the solution proposed. A measure of energy efficiency can be determined and derived from the average bit-error rate within a communication environment. T-UAV$^{RIS}$ improves energy efficiency in one approach where the RIS supports single, full-angle reflection and is mounted on a T-UAV [12]. The aerial RIS can reflect intelligently and improve ground node LoS communication, strengthening the overall channel. The channel is assumed to have dual-hop communication, and the channel state information is known to ensure maximizing the SNR. The results were derived against the average bit-error rate based on modulation order on many passive RIS elements. The results showed that with more passive elements, the average bit error rate degraded and improved EE while providing a reliable environmental communication channel.

In natural disasters, there is an increased demand for mobile data exchange to assist in aiding those in need with minimum delay. UAVs excel at assisting in these approaches due to their flexibility and high degrees of freedom. A fleet of T-UAVs mounted with RISs are deployed at multiple altitudes to serve users requiring mobile data in the emergency area, where the tethered cable increases the energy efficiency of the T-UAV$^{RIS}$ and assists with the flight trajectory of the T-UAV to be within the vicinity of the needed users. To further aid the high demand for data, simultaneous wireless information and power transfer are considered since the RIS elements can be subdivided into collecting energy and reflecting energy-based tasks, which enhances the energy efficiency of the T-UAV$^{RIS}$. The energy efficiencies and channel capacities are proven to increase with integrating T-UAV$^{RIS}$ solutions, especially in high-data demanding disaster areas [34].

Energy efficiency receives a significant boost with T-UAV$^{RIS}$ approaches. T-UAVs ensure a continuous power supply, eliminating the need for frequent battery replacements and recharging. This consistent energy source sustains UAV operations, reducing energy-related disruptions. The integration of RIS technology optimizes signal reflection and redirection, enhancing wireless communication quality while minimizing power wastage. Together, tethered UAVs and RIS create a synergy that extends mission durations and significantly improves energy efficiency in communication systems, making them more sustainable and cost-effective.

## 7.6 T-UAV$^{RIS}$ and U-UAV$^{RIS}$ Case Studies

In the rapidly evolving wireless communications landscape, RIS has emerged as a transformative technology capable of reshaping how we perceive and harness wireless signals. Within this context, our investigation delves into two pioneering case studies: T-UAV$^{RIS}$ and U-UAV$^{RIS}$. The advantage of RIS technology lies in its potential to revolutionize signal propagation and mitigate signal blockages. As our first case study, T-UAV$^{RIS}$ explores the ingenious fusion of RIS technology with T-UAVs, addressing the challenges of urban wireless connectivity. Traditionally, RISs have adorned building facades, presenting limitations related to building owners' willingness to cooperate and signal coverage.

In contrast, as an innovative approach, T-UAV$^{RIS}$ leverages UAVs' agility to optimize LoS connectivity while also addressing inherent battery constraints. Meanwhile, U-UAV$^{RIS}$, our second case study, harnesses the unbridled potential of RIS technology through UAVs that soar freely in the sky. By strategically positioning RIS panels and intelligently manipulating signal propagation, this case study extends the scope of RIS technology, enhancing UAV communication links by mitigating path loss, signal fading, and interference. Together, these case studies illuminate the boundless possibilities and potential challenges that Tethered and Untethered UAV$^{RIS}$ configurations bring to the forefront of wireless communications, providing valuable insights into their real-world applications and contributions to the ever-evolving field of wireless technology.

### 7.6.1 Case Study 1: Fronthaul Outage Compensation in T-UAV$^{RIS}$

**Motivation:** Mounting the RIS panel on an U-UAV will increase the probability of achieving LoS connectivity in addition to allowing 360° panoramic full-angle reflection [17]. Moreover, the U-UAV can adjust the RIS location periodically, according to the user movement, thus maintaining persistent LOS links with both source and destination. However, the U-UAV service time will be limited to its battery capacity (the flying time will be limited to tens of minutes).

These attributes motivated us to present our T-UAV$^{RIS}$ approach, where we consider co-locating a T-UAV$^{RIS}$ to selected small gNBs (SgNBs) in 5G and beyond networks. Given that most SgNBs are fronthauled using wireless links, there is a relatively high probability that any link can fail. Once a fronthaul failure occurs, the T-UAV$^{RIS}$, colocated with the SgNB, will fly vertically to a pre-defined altitude to achieve one or more LoS alternate fronthaul links. If the fronthaul failure occurs to one of the SgNBs that is not equipped with a UAV$^{RIS}$, a neighboring UAV$^{RIS}$ equipped SgNB will launch its UAV$^{RIS}$ to heal the failed SgNB. This fail-safe reflects the importance of the pre-planning phase, where the main objective is to minimize the number of colocated UAV$^{RIS}$ subject to a guaranteed minimum fronthaul achievable rate in case of the failure of any fronthaul link of any SgNB.

**System Design:** We consider a dense network for our fronthaul outage compensation scheme where SgNBs are distributed either at the street level or above buildings in the presence of different-shaped buildings. We consider three different types of SgNBs:

1. Failed SgNB (FSgNB) is an SgNB that is suffering from fronthaul failure, and the proposed scheme is mitigating this failure by using UAV$^{RIS}$.
2. Donor SgNB (DSgNB) is the SgNB that provides the fronthaul connection to the FSgNB before the occurrence of the failure.
3. Neighboring SgNB is a neighboring SgNB to the FSgNB that is going to provide the alternate fronthaul link via UAV$^{RIS}$. For simplicity, we did not show DSgNB in Fig. 7.5.

As shown in Fig. 7.5, the FSgNB is not equipped with our proposed solution, i.e., UAV$^{RIS}$; however, the NSgNB is equipped and launched its UAV$^{RIS}$ to initiate the healing process. Note that the distance between NSgNB and FSgNB is $d$.

We introduce two types of UAVs, T-UAV$^{RIS}$ and U-UAV$^{RIS}$, to show how the orientation of the UAV on the x-axis will affect the achievable fronthaul rate of the FSgNB. Both UAVs are placed at an altitude of $h$, and the U-UAV can move anywhere on the x-axis between NSgNB ($x = 0$) and FSgNB ($x = d$). Given the building that is blocking the LoS communications between FSgNB and NSgNB, both UAVs can achieve connectivity via two LoS links aided by either the T-UAV$^{RIS}$ or the U-UAV$^{RIS}$.

Compared to the RIS mounted on the facades of buildings, T-UAV$^{RIS}$ has multiple appealing advantages. First, with an elevated position, T-UAV$^{RIS}$ can establish LoS links between FSgNB and NSgNB with high probability, which leads to a higher

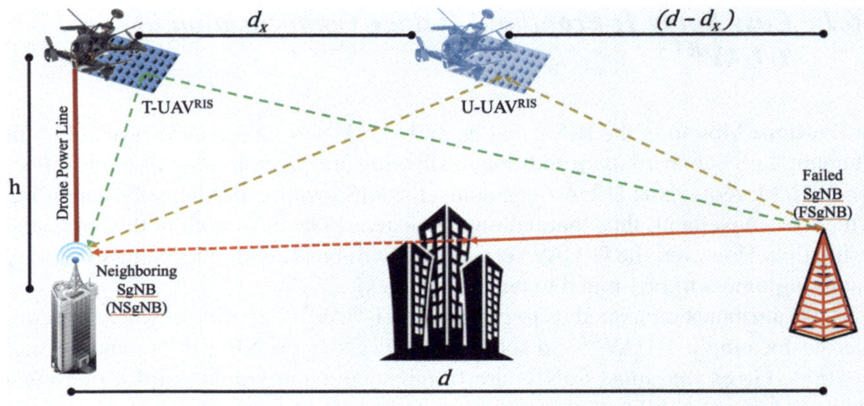


**Fig. 7.5** Fronthaul outage compensation using U/T-UAV$^{RIS}$

fronthaul rate compared to the terrestrial RIS, thereby offering a new degree of freedom to enhance the degree of recovery from failure. Second, equipping one SgNB with T-UAV$^{RIS}$ can heal its fronthaul failure and neighboring SgNB fronthaul failures.

Despite the appealing advantages of the proposed approach, it is still susceptible to some limitations:

- The maximum altitude that the T-UAV$^{RIS}$ can reach is limited to the city/state/federal regulations, which may limit its functionality of achieving a clear LoS between the source and the destination. Also, the optimal tethering location may not be available with some restrictions (for example, privacy) related to flying UAVs near buildings.
- The tethering property will limit the inclination angle of the T-UAV$^{RIS}$, which limits the mobility of the T-UAV$^{RIS}$ to a few meters around the SgNB given that the U-UAV$^{RIS}$ will mainly suffer from the depletion of its battery.

**Simulation Results:** The numerical results simulated using MATLAB are demonstrated in this section to showcase the effectiveness of our proposed approach and explore the advantages of implementing RIS to alleviate X-haul failures in 5G networks. To further investigate details of this simulation's code, refer to [35], where an example of RIS is shown compared to Decode-and-Forward Relaying.

We consider two SgNBs, a U-UAV$^{RIS}$ with $N$ reflecting elements. The locations of the SgNBs and the UAVs are shown in Fig. 7.5. Denote the horizontal distance between the FSgNB and NSgNB by $d$ (100 m in our simulation setup), the distance between the U-UAV$^{RIS}$ and NSgNB as $d_x$, the distance between the U-UAV$^{RIS}$ and FSgNB as $d - d_x$, the altitude of the U-UAV$^{RIS}$ as h. It is assumed that LoS links dominate the NSgNB-IRS and IRS-FSgNB channels, an LoS link dominates the DSgNB-FSgNB channel, and an NLoS link dominates the NSgNB-FSgNB channel.

In this setup, we configure the transmit power to be 17 dBm, the altitude of the U-UAV$^{RIS}$ to be 15 m and 35 m based on the two presented scenarios, the minimum threshold for the fronthaul rate to be 1.5 bits/s/Hz, and the distance between NSgNB and FSgNB is $d$ = 100 m. We used three values for the number of RIS elements: 75, 200, and 500. It is worth noticing that as the number of elements increases, the RIS panel's weight will increase, and the power needed for these elements to operate will increase. In addition, the smart controller's power consumption will increase. All these drawbacks of increasing the number of reflecting elements will not affect our proposed model, given that the tethering property provides a continuous power source.

We modeled the channel gains for the sub-6 GHz (3GHz) using the 3GPP Urban Micro [36]. We used both the LoS and NLoS models. Our analysis considers the channel deterministic, as the RIS elements reflect passively with a fixed phase shift and do not have any signal processing capabilities. The destination, i.e., FSgNB, is supposed to know the channel, and the phase shift can be optimized. This assumption is made due to the non-trivial nature of channel estimation for RIS. However, recent studies such as [37] have presented a general framework for estimating the channel. Additionally, authors in [38] proposed incorporating non-uniformly distributed active elements into the RIS panel to facilitate channel training/estimation through deep learning techniques.

The fronthaul achievable rate in Figs. 7.6 and 7.7 can be calculated using the following three equations:

$$R_{DF}^{LoS} = \log_2\left(1 + \frac{p\beta_{DF}^{LoS}}{\sigma^2}\right) \tag{7.1}$$

$$R_{NF}^{NLoS} = \log_2\left(1 + \frac{p\beta_{NF}^{NLoS}}{\sigma^2}\right) \tag{7.2}$$

$$R_{NIF} = \log_2\left(1 + \frac{p\left(\sqrt{\beta_{NF}^{NLoS}} + \epsilon\, N_x\sqrt{\beta_{NIF}^{LoS}}\right)^2}{\sigma^2}\right) \tag{7.3}$$

where $R_{DF}^{LoS}$ is the achievable rate of the LoS link between DSgNB and FSgNB, $R_{NF}^{NLoS}$ is the achievable rate of the NLoS link between the NSgNB and FSgNB, $R_{NIF}$ is the virtual LoS link from the NSgNB to the RIS then from the RIS to the FSgNB, $p$ is the transmitted power, $\sqrt{\beta}$ is the deterministic flat-fading channel gain, and $N$ is the number of reflecting elements in the RIS panel. We are using the terms channel capacity and achievable rate interchangeably, given that the achievable rate unit is bit/s/Hz.

As shown in Fig. 7.6 where the altitude of the U-UAV$^{RIS}$ is set to be 15 m, the fronthaul achievable rate before the occurrence of failure, i.e., DSgNB-FSgNB LoS

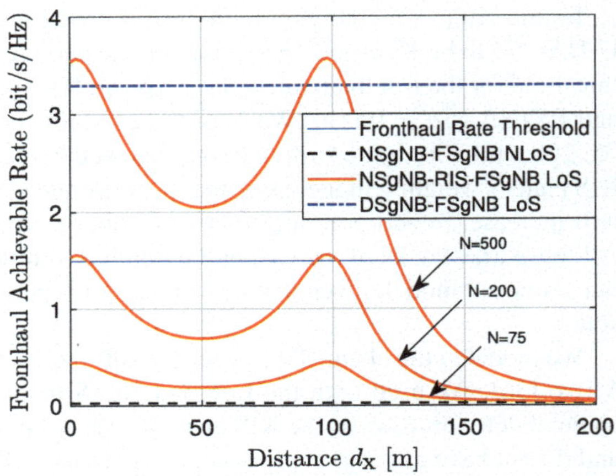


**Fig. 7.6** Fronthaul achievable rate versus distance (h = 15 m)

curve, is around 3.25 bits/s/Hz. Once the failure occurs and before the launching of the U-UAV$^{RIS}$, the NLoS fronthaul link between the NSgNB and the FSgNB, i.e., NSgNB-FSgNB NLoS curve, is very close to 0 bits/s/Hz which shows that without using our proposed solution, the FSgNB will not achieve the minimum required threshold rate. Finally, the three curves bounded by the DSgNB-FSgNB LoS curve and NSgNB-FSgNB NLoS curve represent the fronthaul achievable rate when launching the U-UAV$^{RIS}$ and moving it from $x = 0$ m, i.e., NSgNB location, passing over the FSgNB at $x = 100$ m and continuing until reaching $x = 200$ m. For the three curves that represent different numbers of reflecting elements, it is observed that there are two peaks at $x = 0$ m and $x = 100$ m where the NSgNB and FSgNB are located, respectively. This shows that the optimal location for U-UAV is to hover above either the NSgNB or the FSgNB. Also, for $N = 200$, we can achieve the fronthaul threshold rate at these two locations only. More specifically, when $N = 500$, the fronthaul achievable rate exceeds that of the LoS rate at $x = 0$ m and $x = 100$ m. This occurs under the condition that all reflecting elements are dedicated to reflecting the received signal from the NSgNB to the FSgNB.

To achieve the fronthaul rate threshold and based on the intersection of the fronthaul achievable rate threshold horizontal line and the RIS curves, we have to equip the U-UAV$^{RIS}$ with reflecting elements ranging from 200 to 500 and locate the U-UAV$^{RIS}$ vertically above either the NSgNB, i.e., $x = 0$ m or the FSgNB, i.e., $x = 100$ m. This is the main motive behind proposing our T-UAV$^{RIS}$, where the UAV should be tethered to one of the SgNBs.

It is worth noting that as the U-UAV$^{RIS}$ moves away from the two SgNBs, i.e., $d > 100$ m, the achievable rate degrades rapidly until reaching the NSgNB-FSgNB NLoS rate.

For the same simulation setup and the same parameters except increasing the altitude of the U-UAV$^{RIS}$, h, from 15 to 35 m, Fig. 7.7 shows that increasing the altitude of the UAV degraded the fronthaul achievable rate by more than 50% where even with N = 500 reflecting elements, our proposed approach can achieve the

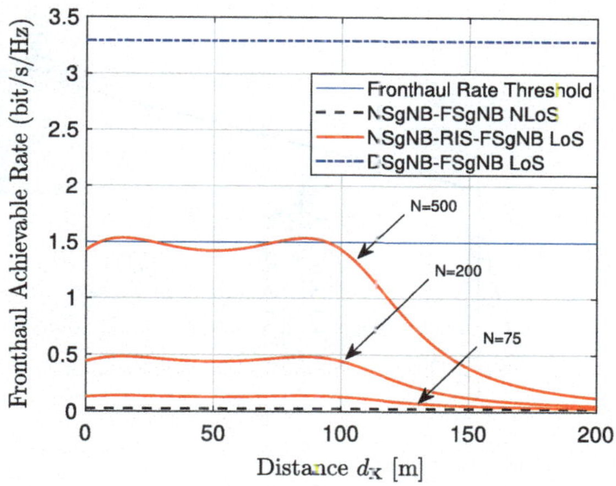


**Fig. 7.7** Fronthaul achievable rate versus distance (h = 35 m)

fronthaul rate threshold only at the two previously proposed locations. To achieve a better fronthaul rate, the source or the destination has to increase the transmit power, or the UAV has to fly at a lower altitude. Moreover, for $N = 200$ and 75 reflecting elements, we lost the advantage of having the UAV tethered to the NSgNB or the FSgNB, where any point on the x-axis will give approximately the same fronthaul achievable rate.

## 7.6.2 *Case Study 2: U-UAV-RIS*

**Motivation:** RIS offers a remarkable capacity for intelligently managing wireless signal propagation, ultimately leading to the establishment of virtual, steady, line-of-sight connections. A primary motivation for embracing RIS technology lies in its capability to tackle the challenges posed by signal blockages in wireless communications. With strategic placements of RIS panels along the signal path, signals can be astutely redirected around obstacles or even harnessed to generate constructive interference. This effectively diminishes the impact of signal blockages, ensuring seamless, and uninterrupted connectivity. RISs bring unique capabilities to enhance U-UAV) communications. RIS technology significantly elevates the quality of U-UAV communication links by intelligently manipulating the wireless propagation environment. RIS elements can combat path loss, signal fading, and interference through precise adjustments to the phase and amplitude of reflected signals. This results in improved signal strength and a reduction in communication errors. The consequence is an expansion of the coverage area, an increase in communication range, and a boost in the overall reliability of U-UAV communications.

**Description of System and Parameters:** The proposed communication system is designed for the distribution of information from a BS to multiple users located in the

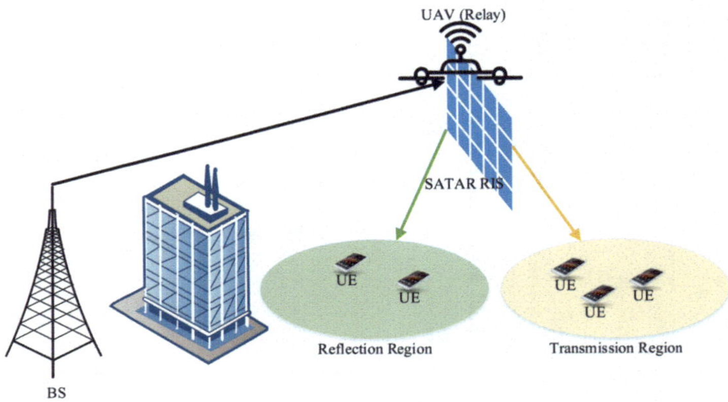


**Fig. 7.8** Proposed system model

transmission and reflection zones of an ASTAR-RIS. In this system, U-UAV plays a pivotal role in establishing connections between the BS and the users, with the ASTAR-RIS being securely mounted on the U-UAV. In this configuration, each user utilizes a single antenna, while the BS boasts a total of Y antennas. The mobility of the U-UAV, capable of hovering over users to deliver services, necessitates consideration of its initial and final positions. It is noteworthy that, as evidenced in [39], the links between the U-UAV and ground nodes may only sometimes adhere to the traditional line-of-sight model. Non-line-of-sight connections can arise due to factors like the U-UAV flying at lower altitudes, leading to potential obstructions from tall structures, trees, and other physical impediments. In the model depicted in Fig. 7.8, the BS efficiently redirects incoming signals toward the ASTAR-RIS mounted on the mobile U-UAV. Subsequently, this technology facilitates connections with the users located on the ground. An implicit assumption in this model is that the U-UAV possesses essential knowledge regarding the locations of both the users and the BS. The ASTAR-RIS has $S$ elements, numbered as $s = 1, 2, ..., S$. We define the set of users as $T$, where $t = 1, 2, ..., T$. The set of users located on the reflection side of the ASTAR-RIS are numbered as $R$, where $r = 1, 2, ..., R$. The proposed model adopts a coordinate system to specify the positions of the BS, the U-UAV, and the transmission/reflection users.

The U-UAV is operational for $L$ seconds at a fixed altitude of $A$. The continuous flight time of the U-UAV, denoted by $l$, is divided into $U$ number of discrete time slots, where $u = 1, 2, ..., U$, of duration $\rho = \frac{L}{U}$ seconds each to simplify calculations. The initial and final locations of the U-UAV are given by $(x[1], y[1])$ and $(x[U], y[U])$, respectively. U-UAV changes its adjacent locations between $(x[1], y[1])$ and $(x[U], y[U])$. $t$ is located at the coordinates $(x_{tu}, y_{tu})$, whereas $r$ is found at $(x_{ru}, y_{ru})$. The BS is positioned at $(x_a, y_a)$. The U-UAV is assumed to possess knowledge of the locations of all users, as well as that of the BS. Table I shows the mathematical symbols used in the paper (Table 7.2).

**Table 7.2** List of mathematical symbols

| Symbol | Description | Symbol | Description |
|---|---|---|---|
| $\mathbf{G_{BS}}$ | BS-RIS gain | $S$ | ASTAR elements |
| $\gamma_i$ | Amplitude | $\boldsymbol{\Theta}_r$, $\boldsymbol{\Theta}_t$ | RIS phase shift |
| $L, U$ | Time | $R$ | Reflection user |
| $\rho$ | Time slots | $\gamma$ | Path loss exponent |
| $(x, y)$ | U-UAV location | $\mathbf{b}_{\{.\}}$ | Beamforming vector |
| $(x_a, y_a)$ | BS location | $T$ | Transmission user |
| $(x_{tu}, y_{tu})$ | $t$ location | $s^n_{\{t,r\}}$ | Received signals |
| $(x_{ru}, y_{ru})$ | $r$ location | $\sigma_e^2$ | Noise |
| $s^n$ | Incident signals | $d_{\{.\}}$ | Nodes distance |
| $\mathbf{g_{st}}$ | $t$-RIS gain | $\omega$ | Variable |
| $\mathbf{g_{sr}}$ | $r$-RIS gain | $j_1,\ j_2$ | Variables |
| $G_{tr}, G_{rc}$ | Gains | $\psi_t,\ \psi_r$ | Constants |
| $p_{\text{base}}$ | BS power | $[.]^f$ | Feasible point |
| $A$ | U-UAV altitude | $[.]^*$ | Optimal point |

**Properties of ASTAR-RIS:** We define the diagonal coefficient matrix of ASTAR-RIS for transmission space user, $t$, as follows:

$$\boldsymbol{\Theta}_t = \text{diag}\left(\gamma_{t,1}e^{j\theta_{t,1}}, \ldots, \gamma_{t,n}e^{j\theta_{t,n}}, \ldots, \gamma_{t,N}e^{j\theta_{t,N}}\right) \tag{7.4}$$

where $\boldsymbol{\Theta}_t \in \mathbb{C}^{S\times S}$. Amplitude and phase shift of the element, $S$, are defined as $1 \leq \gamma_{t,n}$ and $\theta_{t,n} \in [0, 2\pi)$, respectively. We also define the diagonal coefficient matrix of ASTAR-RIS for reflection space user, $r$, as follows:

$$\boldsymbol{\Theta}_r = \text{diag}\left(\gamma_{r,1}e^{j\theta_{r,1}}, \ldots, \gamma_{r,n}e^{j\theta_{r,n}}, \ldots, \gamma_{r,N}e^{j\theta_{r,N}}\right) \tag{7.5}$$

The element, $S$, facilitates amplitude and phase shift adjustments, represented by $\gamma_{r,n}$ and $\theta_{r,n}$, respectively, on the incident signals during reflection. These adjustments are characterized by the complex matrix $\boldsymbol{\Theta}_r \in \mathbb{C}^{S\times S}$, where $1 \leq \gamma_{r,n}$ and $\theta_{r,n} \in [0, 2\pi)$. We define $s^n$ as the wireless signals' incident on the ASTAR-RIS panel. When a wireless signal reaches ASTAR element $S$, the signal is split into transmitted and reflected signals as $s_t^n$ and $s_r^n$, respectively. Recent research in [40] suggests that the reflection and transmission properties of $S$ are generally independent. The transmitted signal by ASTAR-RIS element, $S$, is $s_t^n = s^n\gamma_{t,n}e^{j\theta_{t,n}}$ [41]. Similarly, the reflected signal by ASTAR-RIS, element, $P$ is $p_r^n = p^n\gamma_{r,n}e^{j\theta_{r,n}}$. According to the law of energy conservation, $\gamma_{t,n}$ and $\gamma_{r,n}$ are coupled.

**Channel Gains:** Considering the presence of significant obstacles, it is assumed that there is no direct line-of-sight communication between the BS and the users. In this context, beamforming vectors are introduced at the BS for both the transmission user

($t$) and the reflection user ($r$), represented as $\mathbf{b_t}$ and $\mathbf{b_r}$, respectively. These vectors belong to the complex vector space $\mathbb{C}^{Y\times 1}$. The channel gains in this setup include the link from the BS to the ASTAR-RIS ($\mathbf{G_{BS}} \in \mathbb{C}^{S\times Y}$), the link from the ASTAR-RIS to user $t$ ($\mathbf{g_{st}} \in \mathbb{C}^{1\times S}$), and the link from the ASTAR-RIS to user $r$ ($\mathbf{g_{sr}} \in \mathbb{C}^{1\times S}$). For channel modeling, this paper employs a technique known as block flat-fading. This method divides the communication channel into smaller blocks and assumes that each block experiences flat fading. In addition, large-scale path loss is incorporated into the channel model. Large-scale path loss accounts for the attenuation of the signal as it propagates over a distance. The formula for large-scale path loss considers various factors such as distance and environmental conditions to estimate the signal strength degradation over distance.

**BS to $t$ Data Transmission:** To achieve rate using $\mathbf{g_{st}}\Theta_\mathbf{t}\mathbf{G_{BS}}$, we use the expression:

$$r_t[u] = \log_2\left(1+\frac{p_{\text{base}}|\mathbf{g_{st}}\Theta_\mathbf{t}\mathbf{G_{BS}}\mathbf{b_t}|^2}{\sigma_e^2 d_1^\gamma[u] d_2^\gamma[u]}\right) \tag{7.6}$$

where $d_1[u] = \sqrt{(x[u]-x_a)^2+(y[u]-y_a)^2+A^2}$, $p_{\text{base}} \in \mathbb{R}_+$ denotes the BS transmission power, and $d_2[u] = \sqrt{(x[u]-x_{tu})^2+(y[u]-y_{tu})^2+A^2}$.

**BS to $r$ Data Transmission:** The equation for the data rate between the BS and the transmission user, $r$, is as follows:

$$r_r[u] = \log_2\left(1+\frac{p_{\text{base}}|\mathbf{g_{sr}}\Theta_\mathbf{r}\mathbf{G_{BS}}\mathbf{b_r}|^2}{\sigma_e^2 d_1^\gamma[u] d_3^\gamma[u]}\right) \tag{7.7}$$

where

$$d_3[u] = \sqrt{(x[u]-x_{ru})^2+(y[u]-y_{ru})^2+A^2} \tag{7.8}$$

We jointly optimize the number of ASTAR-RIS phase shifts and U-UAV location to maximize the data rate over $T$ and $R$ number of users subject to the U-UAV mobility constraint.

$$\max_{x[u],y[u],\Theta_\mathbf{r},\Theta_\mathbf{t}} \sum_{u=1}^{U}\left[\sum_{t=1}^{T} r_t[u] + \sum_{r=1}^{R} r_r[u]\right] \tag{7.9a}$$

$$\text{s.t. } (x[u+1]-x[u])^2+(y[u+1]-y[u])^2 \leq \left(\rho v + \frac{1}{2}\rho^2 a\right)^2 \tag{7.9b}$$

$$x[1] = x[N],\ y[1] = y[N] \tag{7.9c}$$

$$r_t[u] = \log_2\left(1+\frac{p_{\text{base}}}{\sigma_e^2}\frac{|\mathbf{g_{st}}\ \Theta_\mathbf{t}\mathbf{G_{BS}}\mathbf{b_t}|^2}{d_1^\gamma[u]\quad d_2^\gamma[u]}\right) \tag{7.9d}$$

$$r_r[u] = \log_2\left(1+\frac{p_{\text{base}}}{\sigma_e^2}\frac{|\mathbf{g_{sr}}\ \Theta_r\mathbf{G_{BS}}\mathbf{b_r}|^2}{d_1^\gamma[u]\quad d_3^\gamma[u]}\right) \tag{7.9e}$$

**Fig. 7.9** Convergence

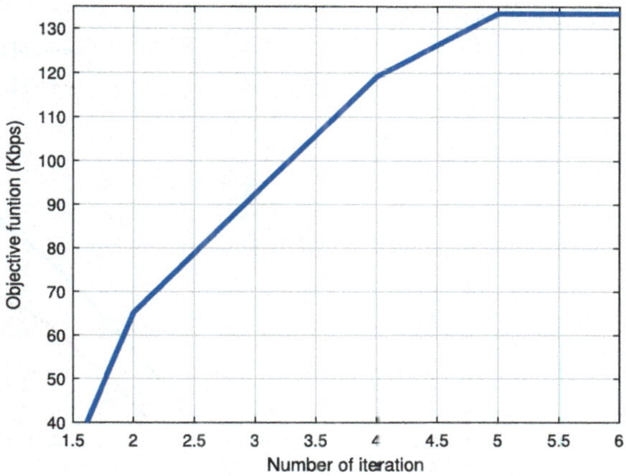


The expressions for $r_t[u]$ and $r_r[u]$ are provided in Eqs. (7.6) and (7.7), respectively. The objective function stated in Eq. (7.9a) characterizes the information-theoretic rate. The constraint on U-UAV mobility, defined in Eq. (7.9b), where $v$ and $a$ correspond to the U-UAV's speed and acceleration, respectively. Equation (7.9c) lays out the constraints related to the U-UAV's initial and final locations. Constraints on the rates of transmission $(t)$ and reception $(r)$ are represented by Eqs. (7.9d) and (7.9e), respectively. The task of finding the optimal solution to the optimization problem presented in Eq. (7.9) is manageable using the GAMS (General Algebraic Modeling System).

**Simulation Results:** Our study presents simulation results in a scenario where $T = R = 2$. We assume that the acceleration and deceleration processes of the ASTAR-RIS mounted on the U-UAV are negligible. The flight time varies within the range of $T \in 5, 10, 15, 20, 25$ minutes, with $\rho = 1$. During this time, the U-UAV hovers from its initial location and returns to the same location. The ASTAR-RIS's altitude is fixed at $H \in 50, 75, 100$ m, and its velocity remains constant at 50 m/s. The BS is situated at coordinates (400, 50) m. The transmission users are positioned at $(0, 0, 0)$ m and $(5, 5, 0)$ m, while the reflection users are located at $(200, 0, 0)$ m and $(200, 5, 0)$ m. In our analysis, we consider a value of $N = 25$. It is essential to acknowledge that an increase in the number of reflecting elements $(S)$ may result in a heavier payload when mounted on a U-UAV. Furthermore, in Fig. 7.9, we offer a visual representation of the relationship between the objective function and the number of iterations for a proposed iterative algorithm. This is specifically illustrated for the scenario when $T = 15$ minutes. The plot reveals that the algorithm converges to a stable solution after the fourth iteration, as evidenced by the flat line on the graph. This observation implies that further iterations do not significantly enhance the algorithm's performance. Consequently, the plot serves as a clear illustration of the algorithm's convergence behavior, aiding in the determination of the optimal number of iterations necessary to attain the desired level of performance.

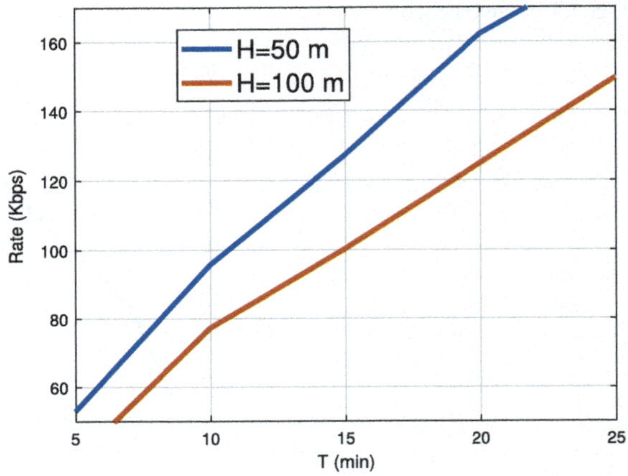


**Fig. 7.10** Data transmission

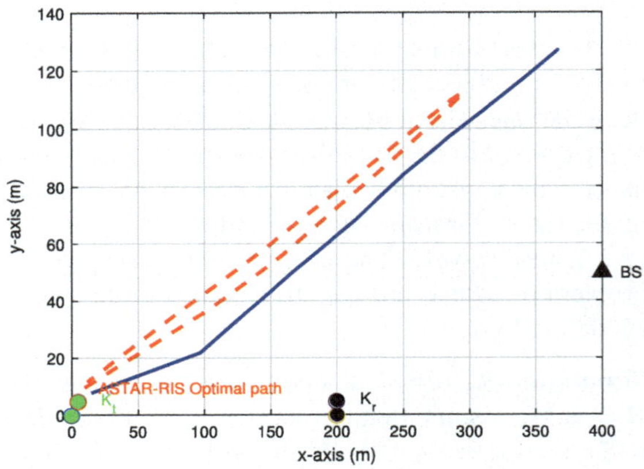


**Fig. 7.11** ASTAR-RIS-assisted U-UAV optimal path planning

The graphical representation in Fig. 7.10 provides a visual depiction of the data transmission rate, measured in Kbps, concerning the number of time slots within an ASTAR-RIS-supported wireless network. Notably, a reduction in the altitude of the ASTAR-RIS corresponds to an increase in the data transmission rate. This observation underscores that the strategic placement of ASTAR-RIS at specific altitudes can lead to a noteworthy enhancement in data transmission rates (Fig. 7.11).

Figure 7.12 presents a visual depiction of the meticulously planned trajectory for U-UAV communications intelligently supported by ASTAR-RIS technology, all within a timeframe of $T = 15$ minutes. In this graphical representation, the horizontal plane defined by the x- and y-axes symbolizes the lateral movement of the U-UAV, while the vertical z-axis delineates the U-UAV's altitude. The visualization meticulously captures the altitude of the U-UAV at various waypoints along its journey, thoughtfully considering the varying heights of the ASTAR RIS, set at 50 m and 100 m. The trajectory featured here is the result of astute optimization, crafted

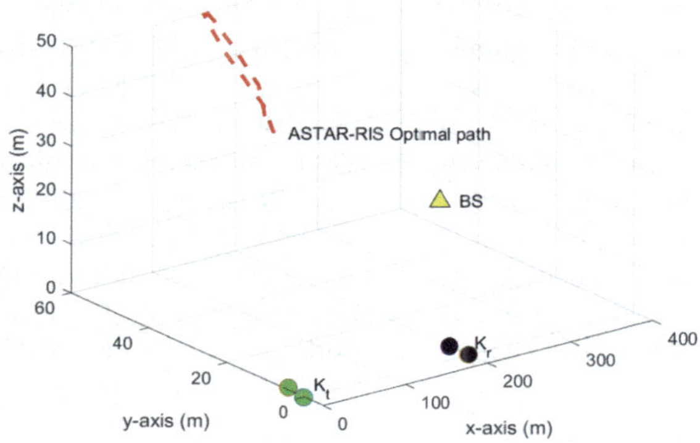


**Fig. 7.12** ASTAR-RIS-assisted U-UAV optimal path planning

to ensure that the U-UAV's movements are efficient in accommodating both transmission and reflection users on the ground, all while elevating the overall Quality of Service. Employing the capabilities of ASTAR-RIS technology, the U-UAV can meticulously adhere to this optimal path. This approach minimizes the distance traveled while maximizing its capability to meet the requirements of both transmission and reflection users. This technological innovation embodies the potential to significantly augment service quality, delivering a noticeably superior user experience.

## 7.7 Conclusions

In conclusion, this chapter has unraveled the exciting potential of merging two cutting-edge domains: Reconfigurable Intelligent Surfaces (RIS) and Unmanned Aerial Vehicles (UAVs). The integration of RIS technology with both tethered and untethered UAVs marks a milestone in wireless connectivity, remote sensing, and network optimization. The synergy between these two fields goes beyond theoretical possibilities, offering practical solutions to complex challenges in modern communication and data delivery. This symbiosis of UAVs and RIS brings solutions to networking and communication challenges in next-generation networks. The adaptability of UAVs, combined with the signal-manipulating capabilities of RIS, results in dynamic trajectory planning, creating Line-of-Sight (LoS) links in previously Non-Line-of-Sight (NLoS) areas. This approach optimizes resource allocation and spectrum sharing, increasing channel capacity and enhancing overall network performance. The marriage of RIS and UAVs offers a promising future for wireless communication, remote sensing, and data delivery. As this partnership continues to evolve, it will reshape the way we connect, communicate, and sense our world, providing innovative solutions to the challenges of an ever-changing wireless landscape.

Demonstrating mounting a UAV with an RIS to mitigate the impact of base station fronthaul failure is seen in the first case study. The approach involves co-locating a tethered or untethered UAV with an RIS to small gNBs. The results suggest that co-locating tethered UAVs with a selected number of base stations can proactively restore any failed fronthaul links. The concept of ASTAR-RISs mounted on UAVs is also introduced in the second case study. ASTAR-RIS possesses the remarkable capability to not only reflect incident signals but also actively amplify them. This innovative approach significantly improves the signal-to-noise ratio (SNR) for multiple users situated in remote or inaccessible areas, expanding the horizons of UAV-assisted wireless networks. Through rigorous simulations, this case study demonstrates the superior network performance and resource allocation efficiency of our proposed ASTAR-UAV model, underscoring its potential to revolutionize the landscape of UAV-assisted wireless networks in the era of 6G and beyond.

## References


1. Tesfaw BA, Juang R-T, Tai L-C, Lin H-P, Tarekegn GB, Nathanael KW (2023) Deep learning-based link quality estimation for ris-assisted uav-enabled wireless communications system. Sensors 23(19). https://www.mdpi.com/1424-8220/23/19/8041
2. Abualhauja'a MO, Ansari S, Popoola OR, Mohjazi L, Bariah L, Muhaidat S, Abbasi QH, Imran MA (2023) Assisted UAV Communications, pp 213–232
3. Rahmatov N, Baek H (2023) Ris-carried uav communication: current research, challenges, and future trends. ICT Express
4. Abdalla AS, Rahman TF, Marojevic V (2020) Uavs with reconfigurable intelligent surfaces: Applications, challenges, and opportunities. https://arxiv.org/abs/2012.04775
5. Ahmed M, Wahid A, Laique SS, Khan WU, Ihsan A, Xu F, Chatzinotas S, Han Z (2023) A survey on star-ris: use cases, recent advances, and future research challenges. IEEE Int Things J 1–1
6. Mu X, Liu Y, Guo L, Lin J, Schober R (2022) Simultaneously transmitting and reflecting (star) ris aided wireless communications. IEEE Trans Wireless Commun 21(5):3083–3098
7. Gupta L, Jain R, Vaszkun G (2016) Survey of important issues in uav communication networks. IEEE Commun Surv & Tutor 18(2):1123–1152
8. Shang B, Shafin R, Liu L (2021) Uav swarm-enabled aerial reconfigurable intelligent surface (saris). IEEE Wireless Commun 28(5):156–163
9. Wang J, Jiang C, Wei Z, Pan C, Zhang H, Ren Y (2019) Joint uav hovering altitude and power control for space-air-ground iot networks. IEEE Int Things J 6(2):1741–1753
10. Selim MY, Kamal AE (2018) Post-disaster 4g/5g network rehabilitation using drones: Solving battery and backhaul issues. In: IEEE Globecom Workshops (GC Wkshps) 2018, pp 1–6
11. Mozaffari M, Saad W, Bennis M, Nam Y-H, Debbah M (2019) A tutorial on uavs for wireless networks: applications, challenges, and open problems. IEEE Commun Surv & Tutor 21(3):2334–2360
12. Kumaravelu VB, Jadhav HK, S AB, Gudla VV, Murugadass A, Imoize AL (2023) Unmanned aerial vehicle-assisted reconfigurable intelligent surface for energy efficient and reliable communication. Springer International Publishing, Cham, pp 173–201. https://doi.org/10.1007/978-3-031-08395-2_8
13. Bushnaq OM, Kishk MA, Celik A, Alouini M-S, Al-Naffouri TY (2021) Optimal deployment of tethered drones for maximum cellular coverage in user clusters. IEEE Trans Wireless Commun 20(3):2092–2108
14. Alzahrani B, Oubbati OS, Barnawi A, Atiquzzaman M, Alghazzawi D (2020) Uav assistance paradigm: state-of-the-art in applications and challenges. J Netw Comput Appl 166:102706. https://www.sciencedirect.com/science/article/pii/S1084804520301806

15. Shi W, Zhou H, Li J, Xu W, Zhang N, Shen X (2018) Drone assisted vehicular networks: architecture, challenges and opportunities. IEEE Network 32(3):130–137
16. Zhang Q, Saad W, Bennis M (2019) Reflections in the sky: millimeter wave communication with uav-carried intelligent reflectors. In: IEEE Global communications conference (GLOBECOM) 2019, pp 1–6
17. Lu H, Zeng Y, Jin S, Zhang R (2020) Enabling panoramic full-angle reflection via aerial intelligent reflecting surface
18. Ndjiongue AR, Ngatched TMN, Dobre OA, Armada AG, Haas H (2021) Analysis of ris-based terrestrial-fso link over g-g turbulence with distance and jitter ratios. J Lightwave Technol 39(21):6746–6758
19. Zhang Z, Dai L, Chen X, Liu C, Yang F, Schober R, Poor HV (2023) Active RIS vs. passive RIS: Which will prevail in 6g? IEEE Trans Commun 71(3):1707–1725. https://doi.org/10.1109%2Ftcomm.2022.3231893
20. Almasoud A, Selim MY, Alsharoa A, Kamal AE (2021) Improvement of bi-directional communications using solar powered reconfigurable intelligent surfaces. In: International conference on computer communications and networks (ICCCN) 2021, pp 1–8
21. Nguyen-Kha H, Nguyen HV, Le MTP, Shin O-S (2022) Joint uav placement and irs phase shift optimization in downlink networks. IEEE Access 10:111 221–111 231
22. Zhou T, Xu K, Shen Z, Xie W, Zhang D, Xu J (2022) Aoa-based positioning for aerial intelligent reflecting surface-aided wireless communications: an angle-domain approach. IEEE Wireless Commun Lett 11(4):761–765
23. Zhu Y, Mao B, Kato N (2022) Intelligent reflecting surface in 6g vehicular communications: a survey. IEEE Open J Veh Technol 3:266–277
24. Tian K, Chi Y, Duo B, Yuan X (2023) Hybrid offline-online design for reconfigurable intelligent surface aided uav communication. IEEE Commun Lett 27(5):1372–1376
25. Cao X, Yang B, Huang C, Yuen C, Renzo MD, Niyato D, Han Z (2021) Reconfigurable intelligent surface-assisted aerial-terrestrial communications via multi-task learning. IEEE J Sel Areas Commun 39(10):3035–3050
26. Mei H, Yang K, Liu Q, Wang K (2022) 3d-trajectory and phase-shift design for ris-assisted uav systems using deep reinforcement learning. IEEE Trans Veh Technol 71(3):3020–3029
27. Eskandari M, Savkin AV, Ni W (2022) Consensus-based autonomous navigation of a team of ris-equipped uavs for los wireless communication with mobile nodes in high-density areas. IEEE Trans Autom Sci Eng 1–13
28. Zhai Z, Dai X, Duo B, Wang X, Yuan X (2022) Energy-efficient uav-mounted ris assisted mobile edge computing. IEEE Wireless Commun Lett 11(12):2507–2511
29. Ren H, Zhang Z, Peng Z, Li L, Pan C (2022) Energy minimization in ris-assisted uav-enabled wireless power transfer systems. IEEE Int Things J 1–1
30. Zhao J, Zhu Y, Mu X, Cai K, Liu Y, Hanzo L (2022) Simultaneously transmitting and reflecting reconfigurable intelligent surface (star-ris) assisted uav communications. IEEE J Sel Areas Commun 40(10):3041–3056
31. Mu X, Liu Y, Guo L, Lin J, Poor HV (2021) Intelligent reflecting surface enhanced multi-uav noma networks. IEEE J Selected Areas Commun 39(10):3051–3066
32. Kishk M, Bader A, Alouini M-S (2020) Aerial base station deployment in 6g cellular networks using tethered drones: the mobility and endurance tradeoff. IEEE Veh Technol Mag 15(4):103–111
33. Khalil AA, Selim MY, Rahman MA (2021) Cure: enabling rf energy harvesting using cell-free massive mimo uavs assisted by ris
34. Yao Z, Cheng W, Zhang W, Zhang T, Zhang H (2022) The rise of uav fleet technologies for emergency wireless communications in harsh environments. IEEE Network 36(4):28–37
35. Bjornson E, Intelligent reflecting surface vs. decode-and-forward: how large surfaces are needed to beat relaying? https://github.com/emilbjornson/IRS-relaying
36. 3rd generation partnership project (3gpp) (2017) Study on channel model for frequencies from 0.5 to 100 ghz, tr 38.901 v14.0.0

37. He Z-Q, Yuan X (2020) Cascaded channel estimation for large intelligent metasurface assisted massive mimo. IEEE Wireless Commun Lett 9(2):210–214
38. Taha A, Alrabeiah M, Alkhateeb A (2019) Deep learning for large intelligent surfaces in millimeter wave and massive mimo systems. In: IEEE global communications conference (GLOBECOM) 2019, pp 1–6
39. Yuan B, He R, Ai B, Chen R, Wang G, Ding J, Zhong Z (2022) A uav-assisted search and localization strategy in non-line-of-sight scenarios. IEEE Int Things J 9(23):23 841–23 851
40. Fu M, Zhou Y, Shi Y (2019) Intelligent reflecting surface for downlink non-orthogonal multiple access networks. In: IEEE globecom workshops (GC Wkshps) 2019, pp 1–6
41. Zhao J, Zhu Y, Mu X, Cai K, Liu Y, Hanzo L (2022) Simultaneously transmitting and reflecting reconfigurable intelligent surface (STAR-RIS) assisted UAV communications. IEEE J Select Areas Commun 40(10):3041–3056

# Chapter 8
# Rayleigh and Ricean Scheme Analysis in Unmanned Aerial Vehicles

**Wasswa Shafik, Ali Tufail, Rosyzie Anna binti Hj Mohd Apong, and Liyanage Chandratilak De Silva**

**Abstract** Considering the Unmanned Aerial Vehicles (UAV) application in communications, the maximum adaption of diversity combination properties has been used for some years facing the Bit Error Rate (BER) of selective combing, maximal ratio, equal gain, direct combining, selective, BER of Rayleigh. and Ricean approach. This study avails a performance evaluation model to analyze the most common technologies in UAV diversity, combining Rayleigh and Ricean based on the diversity combination properties. However, a few models based on the Rayleigh and Ricean schemes have been used for a while in analyzing the communication diversities at a smaller baseline. The existing proposed models have not currently depicted the most reliable and effective model of either adopting Rayleigh or Ricean in UAV communication systems. The study further presents a deep survey of the models and UAV-associated issues with a simplified mathematical expression offered by Rayleigh and Ricean to depict the comparison well. The simulated results demonstrated that the Rayleigh model performs better when subjected to the Ricean model considering the BER.

**Keywords** Drones · Diversity combining · Unmanned aerial vehicle · Rayleigh and Ricean schemes

---

W. Shafik (✉) · A. Tufail · R. A. H. M. Apong · L. C. De Silva
School of Digital Science, University Brunei Darussalam, Gadong, Brunei
e-mail: wasswashafik@dcrlab.org

A. Tufail
e-mail: ali.tufail@ubd.edu.bn

R. A. H. M. Apong
e-mail: rosyzie.apong@ubd.edu.bn

L. C. De Silva
e-mail: liyanage.silva@ubd.edu.bn

W. Shafik
Dig Connectivity Research Laboratory (DCRLab), Kampala, Uganda

© The Author(s), under exclusive license to Springer Nature Singapore Pte Ltd. 2026 
I. Ahmad et al. (eds.), *Reconfigurable Intelligent Surfaces for Unmanned Aerial Vehicle Communication*, Unmanned System Technologies,
https://doi.org/10.1007/978-981-96-9662-8_8

## 8.1 Introduction

Unmanned Aerial Vehicles (UAVs), also known as drones, have significantly transformed multiple industries by providing unparalleled flexibility and ease of use for various purposes [1]. Aerial platforms have produced noteworthy advancements in various domains, including agriculture, by facilitating crop monitoring and precision agriculture [2]. Additionally, they have proven valuable in surveillance and reconnaissance operations, disaster management efforts, and establishing communication networks in distant or demanding settings. Nevertheless, the efficacy of UAVs in these specific use cases is highly contingent upon their capacity to uphold dependable communication connections [3]. Consequently, examining propagation schemes, namely, the Rayleigh and Ricean models, assumes paramount importance.

UAVs possess varying autonomy and are primarily employed in military missions, particularly in remote and inaccessible regions. The utilization of UAVs has expanded across several domains, including demining, civil applications, transportation, commerce, and criminal and terrorist activities, owing to the progress of technology [4]. The study examined two energy management strategies: thermostatic and equivalent consumption minimization. During the learning process, several reinforcement learning (RL) agents make decisions regarding the allocation of resources by employing a weighted average approach. During each cycle iteration, Q-learning agents update their state-action values, with specific details being depicted [5]. The findings indicate that the fuel consumption associated with the proposed comprehensive strategy is 3.2% greater than that of the most optimal individual agent.

In the realm of UAVs, it is imperative to comprehensively comprehend the intricacies associated with wireless communication channels to enhance their operational efficiency effectively. The Rayleigh and Ricean fading models are two essential methodologies employed for modeling the wireless channel characteristics [6]. The Rayleigh fading model is predicated on the assumption of a propagation environment characterized by solely scattered signals, rendering it appropriate for scenarios devoid of prominent signal channels. On the other hand, the Ricean fading model considers the existence of a dominant signal component in addition to scattered signals, providing a more accurate representation of scenarios, including line-of-sight (LoS) propagation [7].

The investigation of fading models is a vital aspect of the analysis of wireless communication channels for UAVs. The Rayleigh and Ricean models are of significant importance within this context. These models offer a theoretical foundation for comprehending the propagation of wireless signals in intricate and ever-changing settings frequently encountered in the context of UAV missions [8]. The Rayleigh fading model is specifically designed to account for the absence of a direct line-of-sight connection between the transmitter and receiver. This makes it particularly applicable when UAVs operate in environments characterized by non-line-of-sight conditions, such as metropolitan areas or mountainous terrains [9].

In contrast, Ricean fading incorporates a constrained line-of-sight element alongside the dispersed signals. The present approach demonstrates its utility in the domain of UAV communication, particularly in scenarios with a partially established but consequential direct signal line [10]. This situation often arises when a drone engages in contact with a ground station that is outfitted with a directional antenna. Comprehending these fading models is of utmost importance as they provide valuable insights into the temporal and spatial variations in signal strength, hence playing a vital role in optimizing UAV communication systems [11].

The examination of Rayleigh and Ricean fading phenomena can significantly influence the operational aspects of UAV systems. Engineers and operators can enhance the design of communication systems by acquiring knowledge about the attributes of the wireless channel, enabling them to develop more resilient systems capable of adapting to dynamic conditions [12]. Adaptability is crucial to maintaining a steady connection between UAVs and their control stations, ensuring the secure and efficient execution of missions. The reliability of UAV communication plays a crucial role in several applications, such as precision agriculture, distant area monitoring, and emergency response [13].

Examining these deteriorating models within the framework of UAVs provides significant perspectives on the characteristics of wireless channels during UAV operations [14]. The acquisition of such insights is of utmost importance in developing communication systems that can adapt to the diverse channel circumstances encountered during UAV operations [15]. By implementing these strategies, operators and engineers of UAVs can enhance the efficiency of data transmission, reduce signal losses, and maintain the UAV's communication capabilities at a high level of performance and reliability during its mission [16].

Also, these fading models' practical implications are particularly relevant when UAVs are deployed in complex settings, such as densely populated urban areas characterized by tall buildings, steep terrains, or thick forests. The wireless connection quality can significantly influence signal reflections, multipath propagation, and signal blockages [17]. The comprehension of how Rayleigh and Ricean fading models apply to these settings facilitates the formulation of communication techniques that effectively minimize interference and improve the dependability of signals [18]. This, in turn, guarantees the successful execution of missions by UAVs and the transmission of crucial data.

The Markov decision process (MDP) is formulated by considering the components of states, actions, and rewards to determine the optimal strategy for many parties regarding path planning and power allocations. A deep learning technique is examined specifically within the RL framework [10]. A UAV with limited energy resources was deployed to collect data from several sensor networks by hovering over them. The study's authors proposed and constructed a learning algorithm to optimize the flight trajectory of the UAV to minimize the regular age-of-the-info of the sensor network while preserving the packet drop rate [13]. Both investigations provided inconclusive evidence about the potential effectiveness of the proposed learning algorithm in significantly reducing the age of information and drop packet rates compared to the baseline greedy algorithms.

In human target search and detection, several optimal approaches are employed to search for and identify the target inside the specified exploration area boundaries. After recognizing the objectives, the UAV can be utilized for both surveillance purposes, capturing visual and auditory data from the targets and returning to its base station either by synchronizing with global positioning system (GPS) components or by transmitting its GPS position to the base station [3]. The MDP is characterized by the components of states, actions, and rewards, which are utilized to determine the optimal strategy for many parties regarding path planning and power distribution. In this study, we investigate the application of deep learning techniques in the context of specialized reinforcement learning. A UAV with limited energy resources was deployed to collect data from each sensor network while hovering over them [19]. The authors of this study have proposed and constructed a learning algorithm to navigate an optimal strategy for the UAV's flight trajectory. This approach minimizes the sensor network's regular age-of-the-info while preserving the packet drop rate. Both experiments have not yet shown conclusive evidence that the proposed learning algorithm can significantly decrease the information age and reduce packet drop rates compared to the baseline greedy algorithms, as illustrated [7].

### *8.1.1 Chapter Key Contributions*

This study presents the following contributions:

- Analyzes the communication diversities and combination properties using UAV standard technologies.
- Avails a performance evaluation model to analyze the technologies in UAV diversity combining Rayleigh and Ricean based on the diversity combination properties.
- A deep literature survey on the state-of-the-art models associated with UAVs, and associated issues.
- A simplified mathematical expression precisely explains the Rayleigh and Ricean models depicted, demonstrating comparison results using BER.

### *8.1.2 The Chapter Structure*

The remainder of this study is arranged as follows. Section 8.2 presents the current technical trends of UAVs. Section 8.3 presents the challenges and limitations of UAVs in communication systems. Section 8.4 details the method of approach within simplified mathematical expressions. Within Sect. 8.5, simulated results are demonstrated and discussed. Section 8.6 holds the present and future works and recommendations. Finally, Sect. 8.7 presents the chapter's conclusion.

## 8.2 Technical Trends of UAV

This section presents current UAV trends, including complex tasking and maneuver decisions, minimum throughput maximization, persistent communication service and fairness, a segmentation method, wireless networking and detection, target tracking, and node positioning.

### *8.2.1 Advanced Sensor Integration*

UAVs are currently experiencing a notable evolution due to incorporating sophisticated sensors. This integration substantially enhances their capacity for gathering and analyzing data [20]. In precision agriculture, UAVs, commonly known as drones, are outfitted with advanced technological components such as high-resolution cameras, multispectral sensors, Light Detection and Ranging (LiDAR) systems, and thermal imaging devices. The sensors acquire substantial data on crops' well-being, the soil's state, and many environmental elements. Through the analysis of this data, farmers can make well-informed decisions about irrigation, fertilization, and pest management [21]. The utilization of precision agriculture techniques facilitates the optimization of crop yields, the minimization of resource utilization, and the reduction of environmental effects.

### *8.2.2 Autonomous Navigation and Artificial Intelligence*

The increasing prevalence of autonomous navigation and the integration of artificial intelligence (AI) in UAVs have significantly enhanced their capabilities, enabling them to function with limited human involvement and increasing their versatility and reliability. The utilization of delivery drones, as demonstrated by companies like Amazon and UPS, serves as a prime illustration of this prevailing pattern. AI algorithms and machine learningof artificial intelligence techniques are utilized to navigate intricate urban landscapes effectively [22]. These algorithms detect and avoid real-time impediments, ensuring the safe and precise execution of deliveries. With the continuous advancement of artificial intelligence, UAVs are increasingly demonstrating enhanced adaptability in many mission scenarios, such as search and rescue operations and environmental monitoring.

### *8.2.3 Swarming Technology*

The utilization of swarming technology is significantly transforming the operations of UAVs by facilitating the seamless collaboration of many drones. UAV swarms are commonly used in military reconnaissance and surveillance operations [23]. These swarms can efficiently and swiftly encompass vast regions, offering military leaders up-to-date information regarding the movements and topography of the opposing forces. Moreover, drones provide redundancy and resilience by enabling individual units to assume responsibilities if others are compromised. In addition to its defensive capabilities, swarming technology exhibits potential in several domains, such as agriculture, wherein the synchronized use of UAVs can effectively survey expansive fields and enhance crop administration [24].

### 8.2.4 Extended Endurance and Range

Recent developments in battery technology and propulsion systems have significantly enhanced UAVs' endurance and range capabilities, creating opportunities for undertaking missions that demand prolonged flight duration and expanded coverage areas [25]. Solar-powered UAVs, shown by the National Aeronautics and Space Administration (NASA) Helios Prototype, utilize solar energy to sustain flight at elevated altitudes for prolonged duration. UAVs are central to various scientific endeavors, such as atmospheric research and climate monitoring [26]. These jobs require extended flight duration to collect extensive data across expansive geographical areas.

### 8.2.5 Vertical Takeoff and Landing (VTOL) Systems

Vertical Takeoff and Landing (VTOL) UAVs possess unique characteristics that amalgamate the benefits of fixed-wing and rotorcraft UAVs, which endows their great versatility. The Parrot Anafi USA is an example of a VTOL UAV that can ascend and descend vertically, obviating conventional runway requirements [27]. This characteristic renders it well suited for tasks conducted in demanding landscapes, such as cartography and topographic analysis in isolated or rugged regions. VTOL UAVs can visit impracticable areas using conventional takeoff and landing techniques.

### 8.2.6 Improved Data Transmission

UAVs depend on effective data transmission to facilitate instantaneous communication and distant data processing. The utilization of advanced technologies, such as high-speed data lines and satellite communication, augments data transmission capabilities. The Matrice 300 RTK by DJI, which incorporates OcuSync Enterprise technology, is a prime illustration of this prevailing tendency [28]. The technology can efficiently transport high-fidelity data across extensive distances while maintaining minimal latency. Incorporating this technology is crucial in scenarios such as live video streaming in search and rescue operations, where the prompt dissemination of information plays a vital role in making informed decisions [29].

### 8.2.7 Counter-UAV Technology

The increasing prevalence of UAVs has generated a corresponding demand for counter-UAV technologies to safeguard vulnerable locations against unauthorized drone incursions. The Anti-UAV Defense System (AUDS) utilizes a combination of radar, RF detection, and jammer functionalities to identify, monitor, and counteract unapproved UAVs [30]. Implementing counter-UAV measures is of utmost importance in protecting vital infrastructure, airports, and public events against potential security risks.

### 8.2.8 Regulatory Compliance and Integration

The continuous development of UAV technology has prompted governments to pursue the establishment of regulatory frameworks actively. These frameworks are designed to facilitate the secure and responsible integration of UAVs into the national airspace [31]. An example of regulatory measures for commercial UAV activities can be observed in the United States, where the Federal Aviation Administration (FAA) has adopted corresponding policies. The UAV requirements pertain to the registration of drones, certification of pilots, and restrictions on airspace [32]. Adherence to these standards is of utmost importance to guarantee the safety and legality of UAV activities while simultaneously promoting advancements within the industry. Table 8.1 demonstrates some trends of UAVs depicting some unique features in the proposed models.

## 8.3 Challenges and Limitations of UAVs in Communications

These additional points underscore the multi-faceted challenges and limitations the UAV industry faces as it continues evolving and expanding its applications. Addressing these issues ensures UAV technology's responsible and sustainable integration into various sectors.

### 8.3.1 Regulatory Hurdles

The intricate and dynamic regulatory framework is one of the foremost obstacles confronting the UAV sector. Governments across the globe are actively engaged in

**Table 8.1** The UAV trending key points and modern UAV models

| Parameter type | UAV entity | Existing works | Reference |
|---|---|---|---|
| Deep reinforcement learning | Autonomous target search | Snake algorithm | [33] |
| Deep learning | Situational awareness | Person-action-locator | [34] |
| An advanced deep neural network | Urban traffic density estimation | Enhanced single shot multi-box detector | [35] |
| Game based | Complex tasking environments | Imitation augmented deep reinforcement learning | [36] |
| Deep Q-network learning | Maneuver decision | DQN training model | [37] |
| Deep Q-learning | Minutest quantity intensification | Various UAV-enabled radio-powered communication net | [38] |
| Deep learning | Persistent communication service, and fairness | Deep deterministic policy gradient | [39] |
| Cloud deep learning | Segmentation methods | A patch-based MVS system | [40] |
| A convolutional neural network | Detection | CRowNet, S-SegNet, and HoughCNet | [41] |
| Q-learning | Wireless networking | Constrained Markov decision process | [42] |

the pursuit of achieving a harmonious equilibrium between the facilitation of innovation and the preservation of safety and security within the realm of airspace [43]. The adherence to rules, encompassing limitations on airspace and requirements for pilot certification, can provide challenges and potentially impede the implementation of UAVs [44]. As an illustration, it is noteworthy to mention that in the United States, the FAA imposes stringent restrictions to govern the operations of UAVs. The operation of UAVs, also known as drones, close to airports or crowded areas, is strictly forbidden without appropriate authorization. Understanding and complying with these restrictions might provide difficulties, especially for enterprises seeking to utilize UAVs for commercial objectives [45].

### *8.3.2 Ranges and Limited Endurance*

Despite recent technological developments, UAVs encounter certain limitations in their ability to sustain flight for extended periods and cover long distances. These limitations can be attributed to the current constraints in battery technology and the overall energy efficiency of UAVs [46]. This constraint restricts the length of missions and the range that UAVs can traverse before necessitating a recharge or refuel. The requirement for regular battery replacements or the need to return to a charging station may impose limitations on long-range surveillance missions or remote infrastructure inspections [46]. Solar-powered UAVs have the potential to increase their operational duration significantly. However, their performance is still limited by the availability of daylight as demonstrated [47].

### *8.3.3 Payload Capacity*

UAVs possess constrained payload capacities, limiting their capacity to accommodate more substantial sensors or equipment. This constraint may impede the implementation of sophisticated sensors or tools necessary for specific applications [48]. In search and rescue missions, the utilization of specialized equipment such as life rafts or medical supplies may be of utmost importance. However, the payload restrictions associated with UAVs can provide significant obstacles [49, 50]. Employing larger drones or exploring alternative delivery techniques may be necessary to solve these limits properly.

### *8.3.4 Limited Autonomous Decision-Making*

Although UAVs are experiencing a rise in autonomy, their capacity to make intricate decisions in dynamic settings remains constrained. Frequently, autonomous aerial vehicles heavily depend on predetermined flight trajectories and may encounter difficulties in promptly adjusting to unanticipated barriers or circumstances [51]. During a firefighting operation, an UAV may encounter dynamic wind patterns or unanticipated obstructions while conducting an aerial assessment of a wildfire [52]. The restricted capacity for autonomous decision-making could impede the entity's capability to adapt its trajectory or acquire crucial information promptly and efficiently [53].

### *8.3.5 Security and Privacy Concerns*

The widespread adoption of UAVs has generated significant apprehension regarding security and privacy problems. The proliferation of drones being employed for surveillance, trespassing, and potentially even terrorist endeavors is an escalating concern [54]. Moreover, acquiring data using UAVs may give rise to privacy apprehensions, particularly in densely populated areas. In 2018, we witnessed delays at Gatwick Airport in the United Kingdom due to the presence of unauthorized drones near the runway [55]. Consequently, this occurrence resulted in the cancellation of flights and substantial economic ramifications. These occurrences highlight the necessity of implementing counter-UAV technologies and implementing strong security measures as depicted [41].

### 8.3.6 Limited Weather Tolerance

UAVs are vulnerable to inclement weather conditions, including but not limited to high wind speeds, precipitation in the form of rain, and reduced visibility caused by fog. The UAV constraints have the potential to adversely affect the dependability and security of missions, particularly in settings characterized by erratic weather phenomena [56]. In the agricultural domain, using UAVs for crop monitoring may encounter hindrances due to unfavorable weather conditions, hence impeding the process of data collecting [57]. Adverse weather conditions, such as heavy rain or strong winds, can impede the safe takeoff and steady flying of UAVs, compromising the quality of data collected and the operation's overall success [58].

### 8.3.7 Data Management and Analysis

UAVs produce substantial quantities of data, presenting a considerable barrier in data management, processing, and analysis. The optimization of data storage, transmission, and processing infrastructure is of paramount importance to fully leverage the value of data collected by UAVs [59]. Environmental monitoring programs utilizing UAVs to acquire high-resolution imagery across extensive regions yield vast amounts of data, often reaching the scale of terabytes. The effort of processing this data on time and converting it into meaningful insights can be intricate and demanding in terms of resources [59].

### 8.3.8 Limited Energy Infrastructure

In regions that are geographically isolated or have experienced natural disasters, UAVs are frequently utilized for emergency relief operations and evaluations. However, these areas may face challenges regarding the restricted availability of energy infrastructure to recharge or refill UAVs [60]. The imposition of this constraint can significantly impede the long-term utilization of UAVs in instances of utmost importance. During disaster response missions, the operational effectiveness of UAVs engaged in communication relay or aerial surveys may be impeded due to the limited availability of energy sources for recharging [61].

### 8.3.9 Air Traffic Management Integration

Incorporating UAVs into current air traffic control systems poses a substantial obstacle. Establishing effective communication and collision avoidance protocols is crucial for ensuring the secure coexistence of UAVs and manned aircraft. Inadequate integration of UAVs into the airspace can result in mid-air collisions or disruptions to manned flight operations [62].

### 8.3.10 Environmental Impact and Noise Pollution

UAVs, especially those equipped with conventional propulsion mechanisms, can potentially contribute to environmental damage and noise pollution. UAVs that rely on fuel combustion as their power source have been shown to release significant amounts of greenhouse gases into the atmosphere [63]. Additionally, the noise these UAVs make during flight operations has the potential to disturb nearby ecosystems and communities. The persistent challenge lies in achieving a harmonious equilibrium between leveraging the advantages of UAV technology and minimizing its adverse effects on the environment and noise pollution, especially in ecologically sensitive regions and densely populated urban areas [64]. Several challenges have been identified with UAVs, which are summarized in Table 8.2.

## 8.4 Methodology

This section presents the mathematical expression of the suggested model with the following notation. $\xi_i$ depicts the instance signal-to-noise ratio (SNR) of the branch$i$, $\phi'$ shows the SNR of combiner output given the $\forall$ explicitly shows the absolute output. DS, with its primary motivation of refining the dependability of data (message) signals using minimally dual, or it may request for more casting (communication) channels with diverse traits. These diversities principally jumble-sale in radio communications, and some of the techniques for contending are fading, co-channel interference, and avoiding error bursts. Some DS options do include the maximum ratio combining (MRC) that mainly ensures that optimum branch weighting through the $X_i = \lambda_i^* = \lambda_i e^{r\phi q}$ to maximize the SNR first. Table 8.3 provides detailed notation and parameters.

The equal gain combination is also another way to scheme out some communication issues where the branch weighs through $X_i e^{r\phi q}$. The $\forall$ depicts the selection diversity exploiting a single branch signal, simplifying the execution of $\forall$ on a single receiver branch. Considering a brute force analysis of MRC ($\phi$) with signal noise observations of $X\,[k]$.

**Table 8.2** A summarized challenge of UAV and state-of-the-art existing work models

| Challenges | Existing works | Results | Loopholes | Reference |
|---|---|---|---|---|
| Image multi-labeling | Neural network | Increased interpretability | Time complexities are considered | [65] |
| Urban traffic density estimation | Random sampling DNN | Experiment replay | Only "And" was used | [66] |
| Situational awareness system | Person-action-locator | Solved multi-UAV cooperation trajectory design | Low communications scenarios were used | [67] |
| Tasking | Imitation augmented deep RL model | Fine-Tuning | Low communications scenarios were used | [68] |
| Maneuver decision | MDP | Increased effectiveness, increased efficiency | Only conventional algorithms were compared too | [69] |
| Minimum throughput maximum | A proximal policy optimization | MSR-PPO shows improvement, RT-PPO depicted efficiency in general performance | Only deep reinforcement learning was used | [70] |
| Persistent communication services fairness | UC-DDPG model | Convergence and output forms | Increased data volume, and increased energy | [71] |
| Segmentation method | Patch-based MVS algorithm | Availed forestry survey tools | Low coverage rates | [71] |
| Crop row detection | CNN | Robust performance | No one model was compared | [72] |
| Power transfer | MDP | Low packet low | Over 39.8% was trusted compared to greedy learning | [71] |
| Efficient animal detection | CNN | Increased identification of animals to over 80 % | Low baseline | [73] |
| Target tracking | A course to the fine deep scheme | Increased accuracy gains | Low ratio of UAV tracking variation | [74] |
| Node positioning | Deep Q-network | Efficiency | Limited position models | [75] |
| Beam selection | Three-dimensional fast machine learning algorithm | optimized utilization of biomedical engineering technology systems | More realistic medical situations are desired in the comparisons | [76] |

**Table 8.3** Notation and parameters

| Notation | Parameter description |
|---|---|
| $\xi_i$ | Depicts instance SNR of the branch $i$ |
| $\phi$ | Depicts the maximum ratio combining |
| $\forall$ | Depicts the absolute output given an instance $i$ |
| $\xi_a v$ | Depicts average signal-to-noise ratio at all branches |
| $h$ | Depicts a complex constant |
| $h_i = \lambda_i e^r \phi q$ | Depicts the channel gain of the branch $i$ |
| $X_i$ | Compound weights for the branch $i$in the output combining |
| $P$ | Depicts the probability |
| $n$ | Depicts an independent white Gaussian noise |
| $x$ | Depicts an arbitrary complex value |
| $\xi$ | Depicts the instantaneous SNR |
| $c$ | Depicts compound constants |
| $C$ | Depicts the link capacity |
| $B$ | Depicts the receiver bandwidth |
| $cdf$ | Cumulative distribution function |

$$y_1[k] = h_1 x[k] + n_1[k], \tag{8.1}$$

$$y_2[k] = h_2 x[k] + n_2[k] \; . \tag{8.2}$$

Given that consider $h_1$and $h_2$are the known compound constants, $n_1$ and $n_2$are independent white Gaussian noise processes with estimation of the original signal expressed in Eqs. 8.1 and 8.2.

$$\hat{x}[k] = h_2 y_1[k] + h_2 y_2[k] \; . \tag{8.3}$$

This approach adheres to regulating the optimal valves of some gains that the signal-to-noise ratio is maximized at the combiner output, for instance, equaling one, which implies that the variance of the expected signal gain as expressed in Eq. 8.3 is one to the variance of $x$; estimation tends to unbiased $x_1 y_1 - x_2 y_1$, and $x_2 = \frac{1 - x_1 y_1}{h_1}$. The noise power is simplified and then given to be

$$\sigma_{\hat{n}}^2 = \left(|x_1|^2 + |x_2|^2\right) \sigma_n^2 \left[ |x|^2 + \left| \frac{1 - x_1 h_1}{\dot{n}_2} \right|^2 \right],$$

$$\sigma_n^2 = \frac{|x_1|^2 + |x_2|^2 + (1 - x_1 y_1)\left(1 - x_1^* y_1^*\right)}{|h_2|^2},$$

$$\sigma_{\hat{n}}^2 = \frac{|x_1|^2\left(|h_1|^2+|h_2|^2\right)+1-x_1h_1-x_1^*y_1^*}{|h_2|^2}\sigma_n^2. \tag{8.4}$$

This is possible when the signal-to-noise ratio is extended with the noise power as simplified Eq. 8.4.

$$\partial = \frac{|x_1|^2\left(|h_1|^2+|h_2|^2\right)+1-x_1h_1-x_1^*y_1^*}{\partial x_1},$$

$$\left(|h_1|^2+|h_2|^2\right)x_1^*-h_1\ . \tag{8.5}$$

The expression depicted in Eq. 8.5 zeros to exactly 0, this leads to the fact that the $x_1$, and $x_2$, respectively.

$$x_1 = \frac{h_1^*}{\left(|h_1|^2+|h_2|^2\right)}x_2 = \frac{h_2^*}{\left(|h_1|^2+|h_2|^2\right)},$$

The signal-to-noise ratio is maximized with the mean squared error (MSE) correspondence given by the following expression:

$$SNR_{\max} = \left(|h_1|^2+|h_2|^2\right)\frac{\sigma_x^2}{\sigma_{\hat{n}}^2}, \text{ and the} J_{\min}\frac{\sigma_{\hat{n}}^2}{\left(|h_1|^2+|h_2|^2\right)}.$$

The signal phase rotation is compensated, and the two observations add up connectively and constructively. This generally means that scaling does not affect the signal-to-noise ratio, but it affects MSE. Therefore, considering the signal gain takes an arbitrary consideration with the complex value being $x$ expressed within Eq. 8.6:

$$|1-x|^2\sigma_x^2+\left(|x_1|^2+|x_2|^2\right)\sigma_n^2, \tag{8.6}$$

Given that$x_1h_1+x_2h_2=x$, and$x_2=\frac{x-x_1h_1}{h_2}$ considering Eq. 8.6. For simplicity in derivation, the noise power is converted into:

$$\sigma_{\hat{n}}^2\left(|x_1|^2+|x_2|^2\right)\sigma_n^2 = \left(|x_1|^2+\left|\frac{x-x_1h_1}{h_2}\right|\right)\sigma_n^2,$$

$$\frac{|x_1|^2+|x_2|^2+(x-x_1h_1)\left(x^*-x_1^*h_1^*\right)}{|h_2|^2},$$

$$\frac{|x_1|^2\left(|h_1|^2+|h_2|^2\right)+|x|^2-x^*x_1h_1-xx_i^*h_1^*}{|h_2|^2}\sigma_n^2 \,. \tag{8.7}$$

For the fixed $x$ from Eq. 8.7, the noise power is minimized as

$$\frac{\partial\left(|x_1|^2+|h_1|^2+|h_2|^2\right)+|x|^2-x^*x_1h_1-xx_i^*h_1^*}{\partial h_1}\sigma_n^2. \tag{8.8}$$

From Eq. 8.8, it signifies connectivity equates to 0, given that $x_1=\frac{xh_1^*}{\left(|h_1|^2+|h_2|^2\right)}$, and the

$$x_2=\frac{xh_2^*}{\left(|h_1|^2+|h_2|^2\right)}.$$

$\sigma_{\hat{n}\min}^2=\frac{|x|^2}{\left(|h_1|^2+|h_2|^2\right)}\sigma_n^2$, the MSE is now to be given the following expression:

$$J=|1-x|^2\sigma_x^2+\frac{|x|^2}{\left(|h_1|^2+|h_2|^2\right)}\sigma_n^2,$$

$$\frac{\partial J}{\partial x}=\left(x^*-1\right)\sigma_x^2+\frac{x^*}{\left(|h_1|^2+|h_2|^2\right)}\sigma_n^2 \,. \tag{8.9}$$

This expression illustrated in Eq. 8.9 demonstrates the returns to 0, thus leading to

$x=\frac{1}{1+\frac{\sigma_{\tilde{x}}^2}{\left(|h_1|^2+|h_2|^2\right)\sigma_{\tilde{x}}^2}}$ and the $S^r J_{\min}=\frac{\sigma_n^2}{\left(|h_1|^2+|h_2|^2\right)+\frac{\sigma_{\tilde{n}}^2}{\sigma_{\tilde{x}}^2}}$.

It mathematically demonstrated that MSE analysis takes some minimum effect on the baseband signal of the diverse diversities (see Eq. 8.9):

$$r_i(t)=h_ix(t)+n_i(t) \,. \tag{8.10}$$

Given that $n_i(t)$ are autonomous compounds with equal variances, the linear combiner output can be articulated as conferring to Eq. 8.10.

$$y(t)\sum_{i=1}^{L}x_i\left(h_ix(t)+n_i(t)\right)+\sum_{i=1}^{L}x_ih_i+\sum_{i=1}^{L}x_in_i(t) \,. \tag{8.11}$$

The instantons signal and noise power at the combiner output in conjunction with Eq. 8.11:

$$\sigma_y^2 = E\left[|x(t)|^2\right].\left|\sum_{i=1}^{l} x_i h_i\right|^2 \text{ leading to } \sigma_{nc}^2 = \sigma_n^2.\sum_{i=1}^{L} |x_i|^2.$$

Therefore, the instantons signal-to-noise ratio output can be taken as

$$y_c = \xi_{av}.\frac{\left|\sum_{i=1}^{L} x_i h_i\right|^2}{\left|\sum_{i=1}^{L} |x_i|\right|^2}.$$

Considering the Schwarz inequality for the parameter, then it is expressed as

$$\left|\sum_{i=1}^{L} x_i h_i\right|^2 \leq \sum_{i=1}^{L} |x_i|^2.\sum_{i=1}^{L} |h_i|^2. \tag{8.12}$$

This maintains the confirmation of $x_i = ch_i^*$ considering that $c$is a compound constant. Considering further that $c = 1$, the instant signal-to-noise ratio is expanded with $x_i = h_i^*$ of Eq. 8.12 and it resulted in

$$\phi = \xi_{av}.\sum_{i=1}^{L} |h_i|^2 = \sum_{i=1}^{L} \xi_i . \tag{8.13}$$

Considering that $\xi_i$ is the instant signal-to-noise ratio of the branch $i$. To achieve the probability distribution function (pdf) of the signal-to-noise ratio at the combiner output. The pdf is obtained in Eq. 8.13.

$$pdf = f_\tau = \frac{1}{(L-1)}.\frac{\phi^{L-1}}{\xi_{av}^L}.e^{-\phi/\xi_{av}}. \tag{8.14}$$

Considering Eq. 8.14 the diverse number of divers $N_r = L$e branches. Most importantly, the outage probabilities can be attained from the cdf as

$$cdf = P(\phi \prec \xi) \int_0^{\xi} f_\tau(\xi\phi)\, d\phi . \tag{8.15}$$

The expression demonstrated in Eq. 8.15 leads to

$$1 - \int_{\xi}^{\infty} f_\tau(d)\,\phi = 1 - e^{-\phi/\xi av} \sum_{i=1}^{L} \frac{(\phi/\xi_{av})^{i-1}}{(i-1)!}.$$

When it comes to signal-to-noise ratio at the ∀ output is

$$\forall = Max\,\{\xi 1, \xi 2, \xi 3, \ldots., \ldots, \xi_{\forall L}\}\ . \tag{8.16}$$

The pdf of the signal-to-noise ratios for the entire branches following the distributions considering Eq. 8.16; say exponentially.

$$f_{\tau}\left(\xi_i\right) = 1/\xi_{av}.\, e\frac{\xi_i}{\xi_{av}}, \qquad\qquad \xi_i \geq 0\,. \tag{8.17}$$

In the reflection to Eq. 8.17, the cumulative distribution is presented as

$$P\left(\xi_i \prec \forall\right) = \int_{-\infty}^{\forall} f_{\tau}\left(\xi_i\right) d\xi_i = 1 - e^{-\forall/\xi_{av}} \qquad \xi \geq 0\,. \tag{8.18}$$

Observing the expression behavior of Eq. 8.18, the cdf for the signal-to-noise ratios after all selection combining can be written as

$$f_{\tau}\left(\forall\right) = P\left(\xi_i \leq \forall,\ i = 1, \ldots., \ldots.L\right), \forall \prod_{i=1}^{L} P\left(\xi_i \leq \forall\right) = \prod_{i=1}^{L}\left(1 - e^{-\forall/\xi_{av}}\right),$$

$$\left(1 - e^{-\forall/\xi_{av}}\right)^{L}, \qquad\qquad \forall \geq 0. \tag{8.19}$$

The outage possibilities in selection combining, the corresponding pdf can be obtained as the derivative of the previous expression contrary to Eq. 8.19.

$$f_{\tau}\left(\forall\right) = \frac{d}{d\forall} f_{\tau}\left(\forall\right), \frac{L}{\forall} 1 - e^{-\forall/\xi_{av}}.\left(1 - e^{-\forall/\xi_{av}}\right)^{L-1}, \qquad \forall \geq 0. \tag{8.20}$$

Conclusively, in Eq. 8.20, the outage capacity of Shannon capacity of the transmitter is Eq. 8.21 expressed as

$$C_s = \int_{0}^{\infty} B \log_2\left(1 + \xi\right)\ P\left(\xi\right) d\xi\,.$$

Given that $C$ illustrates the link capacity, $B$ is the receiver bandwidth, $P$ is the probability, $\xi$ and lastly the instantaneous SNR. This implies that the capacity can be converted as

$$C_{out} = B\left(1 - P_{out}\right) \log_2\left(1 + \forall_{\min}\right). \tag{8.21}$$

## 8.5 Results and Discussion

This section presents the implementation of the proposed diversity combining strategy, which is subjected to the existing model. The simulation requirements include the diverse branches of uncorrelated Rayleigh or Ricean fading signals.

Receiver diversity refers to the form of spatial diversity in which the receiver has multiple antennas. The presence of receiver diversity presents an interesting phenomenon, effectively using information from all UAVs to demodulate data; see Figs. 8.1 and 8.2. The average symbol energy-to-noise power ratio for each branch is 1, 3, 5, 7, and is shown as 9 dB. SNR continues to increase as the number of receivers increases. It also shows that the UAV communication system works better when the spatial-temporal block coding technique is used to receive the diversity. In contrast to BER, the Rayleigh model performed better than the Ricean based on the demonstrated results using UAV as a case study, as shown in Figs. 8.2 and 8.3.

The Equal Gain Combining (EGC) can function within the BER that the multi-cathode denitrification ratings and the hydraulic retention times were considered per the cdf and pdf of the BER of the proposed model as truncated as hydraulic retention times equaling some addition time complexes, often maximized by using the MRC provided that the channel is perfectly estimated at the receiver. The denitrification ratings can be enhanced 3 to 30 times in judgment with the former Ricean model. The EGC successfully presented better performance when subjected to the BER and the SNR, so the run-off attentiveness in the experiment was focused on the SC, and EGC of one through four branches, as depicted in Fig. 8.4.

With the typical BER for Ricean fading channels, the outage perspective tends to descend closed-form subordinate limits for the outage probabilities, and the regular BER might be tight for diverse diminishing circumstances. Additionally, the

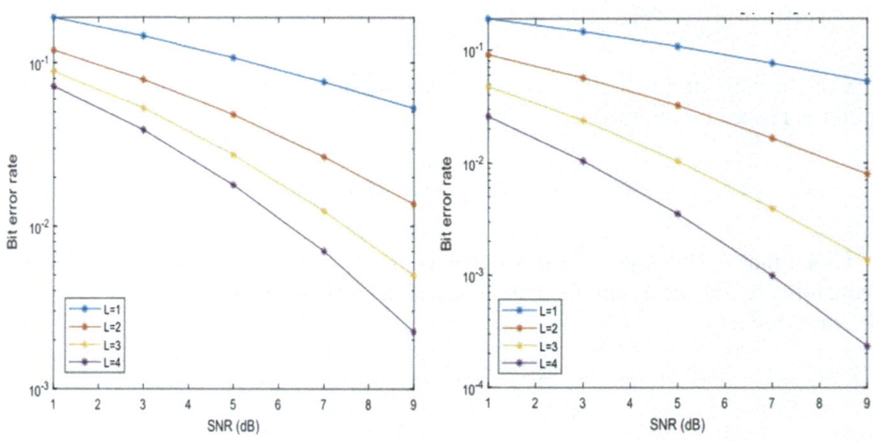


**Fig. 8.1** BER of selective combing of the Rayleigh model (*left panel*) and BER of maximal ratio combining of the Rayleigh model (*right panel*)

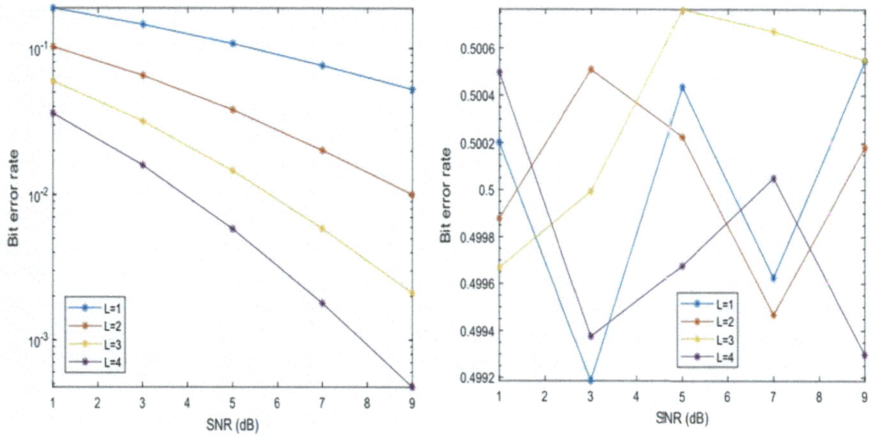


**Fig. 8.2** BER of equal gain combining of the Rayleigh model (*left panel*) and BER of direct combining of the Rayleigh model (*right panel*)

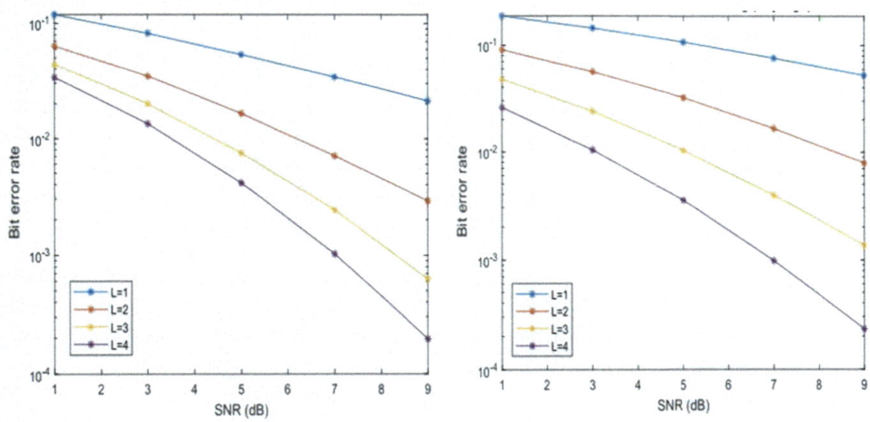


**Fig. 8.3** BER of selective combining of the Ricean model (*left panel*) and BER of the maximal ratio combining of the Ricean model (*right panel*)

model presented and depicted that the asymptotic examination of the outage probabilities and the normal BER gave that it is in the SNR region. In observation, it is also demonstrated that DC (Ricean) can achieve the occupied diversities directive in dual-way communication schemes with appropriate substituting verges. These techniques of SC and MRC exploit anti-cable large-scaled fades triggered through surveillance, highlighting the discrepancies in the topography profiles and the layout of the environs. The SNR enhancement obtainable by SD is straightforwardly evident in that the regular SNR continuously upsurges with an intensification in the number

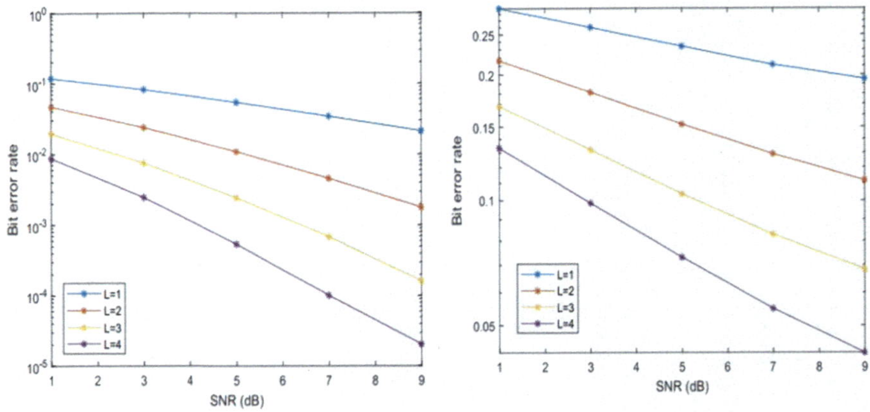


**Fig. 8.4** BER of equal gain combining of the Ricean model (*left panel*) and BER of direct combining of the Ricean model (*right panel*)

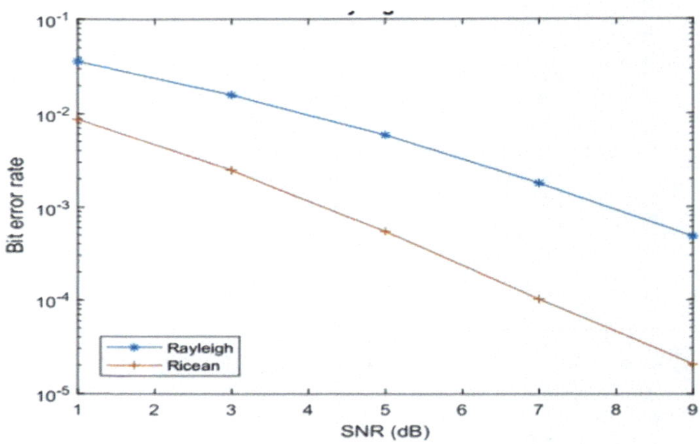


**Fig. 8.5** BER of Rayleigh and Ricean models

of branches. The major variety gain is attained using the two branch diversities, as presented in Fig. 8.5.

However, the democratization of drone technology has led to its use by unfavorable elements of society. The recent spate of attacks by terrorists using commercial drones has led to the need to develop anti-drone equipment. With civilian installations becoming vulnerable to drone-related disturbances, local law enforcement can be expected to deploy anti-drone technology in the future.

The demonstration proves the diverse options of Rayleigh are powerful when it comes to the communication receiver method that affords radio communication connection enhancement at comparatively truncated costs. These simulated techniques

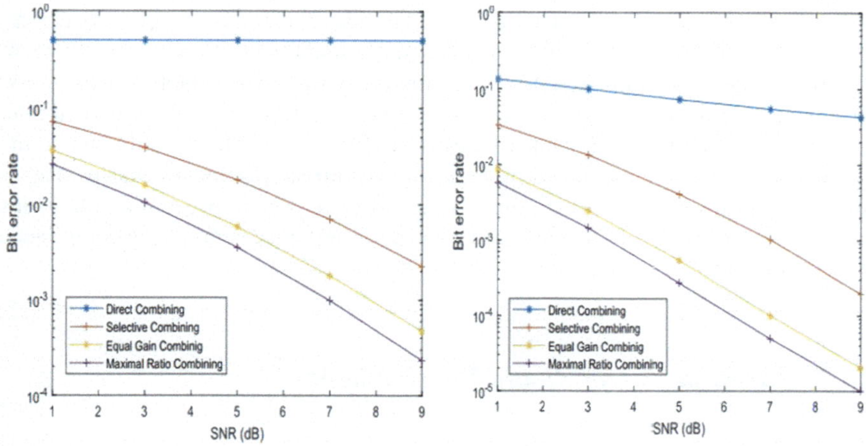


**Fig. 8.6** BER of SC MRC EGC DC of the Rayleigh model (*left panel*) and BER of SC MRC EGC DC of the Ricean model (*right panel*)

tend to feature small-scaled fades that are categorized through cavernous and speedy breadth variations as per moveable transfers ended detachments of unbiassed a few wavelengths availed by the MRC and EGC. The RC performs slickly poorly as compared to the SC and EGC, in comparison to the MRC. This technically implies that within the Ricean category, MRC outcompetes EGC, SC, and DC in these regards, given the BER. It is possible to get better results with wireless communication by following self-regulating signal trails, getting multiple copies of the signal being sent, and then picking one (or more) of them wisely. This, along with the current and normal SNRs at the phone, can sometimes lead to improvements of up to 20 to 30 decibels, as demonstrated in Fig. 8.6.

## 8.6 Future Trends and Recommendations

This section presents some identified trends and recommendations; the earlier identified challenges and limitations in the UAV industry demand a multi-faceted approach involving collaboration, innovation, regulation, and ethical considerations, some as developed.

### *8.6.1 Streamline Regulatory Frameworks*

It is imperative for governments to proactively collaborate with industry stakeholders to establish all-encompassing and uniform regulatory frameworks that effectively cater to the wide array of UAV activities. The laws ought to provide explicit directives for operators of UAVs, guarantee safety measures, and foster advancements in technology. One potential approach to addressing this issue is the establishment of distinct legal classifications for various sorts of UAV operations, including recreational, commercial, and public service activities [68]. This stratification would enable the customization of regulations and prerequisites to suit specific use scenarios. Regulatory authorities must prioritize the advancement of streamlined online platforms that facilitate the acquisition of approvals and permits, hence simplifying compliance procedures.

### *8.6.2 Invest in Research and Development*

Governments, private enterprises, and academic institutes must spend significant resources on research and development endeavors to address the constraints associated with UAVs. It is imperative that funding be allocated toward promoting breakthroughs in battery technology, encompassing the research and development of batteries that possess high-capacity, lightweight characteristics, and exhibit prolonged durability [69]. Furthermore, it is imperative to direct research endeavors toward augmenting the autonomy of UAVs by leveraging machine learning and AI algorithms [70]. This would enable drones to autonomously make instantaneous judgments by analyzing ambient circumstances and sensor information. The advancement and widespread adoption of sustainable propulsion technologies, such as electric or hybrid systems, can significantly enhance UAVs' endurance capabilities.

### *8.6.3 Encourage Industry Collaboration*

Establishing a collaborative environment among producers, operators, and regulatory authorities of UAVs is vital to cultivating a conducive ecology. Public–private partnerships have the potential to collaboratively address issues about security, integration of airspace, and the advancement of UAV technologies [72]. Through the facilitation of open communication and the sharing of knowledge, stakeholders have the ability to collaboratively discover equitable resolutions to intricate challenges, thereby guaranteeing the responsible advancement of the UAV manufacturing sector.

### *8.6.4 Develop Weather-Resilient UAVs*

UAV producers must emphasize the advancement of weather-resistant UAVs that can function effectively in diverse environmental circumstances. This involves the development of UAVs equipped with sophisticated stabilization systems and resilient propulsion mechanisms capable of enduring harsh wind and unfavorable weather conditions. The integration of waterproofing and durability should be considered in the design of UAVs [71]. Weather-resistant UAVs of this nature would possess immense value in various domains, such as search and rescue operations conducted in arduous weather conditions and emergency response missions executed under unfavorable weather circumstances.

### *8.6.5 Enhance Data Management Solutions*

The UAV sector must allocate resources toward implementing comprehensive data management solutions to effectively handle the extensive and varied data created throughout operational activities. This encompasses the advancement of cloud-based storage solutions that are both secure and scalable, enabling efficient management of extensive datasets [73]. To facilitate the prompt and efficient transfer of data, particularly in the context of remote or long-distance missions, it is imperative to utilize resilient data transmission methods. Sophisticated data analytics technologies, including machine learning and artificial intelligence, can accelerate data processing, transforming unprocessed data into practical and usable insights [74]. To uphold the integrity and confidentiality of data, stakeholders must adhere to rigorous rules and guidelines for data security.

### *8.6.6 Address Environmental Concerns*

UAV makers must place utmost importance on advancing environmentally friendly UAVs, by adopting electric or hybrid propulsion systems that effectively mitigate pollutants. Further investigation should be conducted into noise pollution mitigation methods, including developing quieter propulsion systems and exploring new blade designs. These measures are essential to minimize disruptions to ecosystems and local communities [77]. It is recommended that operators proactively seek authorization and follow established rules to minimize disturbances caused by UAVs in environmentally sensitive areas. This approach ensures compliance with local regulations governing UAV operations.

### *8.6.7 Promote Energy Infrastructure in Remote Areas*

It is recommended that governments, in conjunction with humanitarian groups, allocate resources toward developing renewable energy infrastructure, with a specific focus on establishing solar charging stations in geographically isolated or vulnerable areas [78]. The charging stations can furnish a dependable and consistent supply of energy, hence facilitating the sustenance of prolonged UAV operations in crucial scenarios, including but not limited to endeavors about disaster relief initiatives and remote surveys [75]. Implementing such infrastructure will guarantee the continuous functionality of UAVs during critical periods, facilitate effective data acquisition, and expedite disaster management.

### *8.6.8 Develop Air Traffic Management Solutions*

Establishing partnerships among UAV manufacturers, air traffic control authorities, and technology suppliers is imperative to foster the advancement of resilient air traffic management solutions. The proposed solutions should incorporate real-time communication and collision avoidance systems, utilizing new technologies such as 5G and powerful radar systems [76]. Creating protocols that provide smooth coordination between human and unmanned aircraft is imperative. These protocols should encompass well-defined procedures that enable UAV operators to get the requisite permissions for accessing airspace [79]. It is imperative for regulatory agencies to proactively monitor and modify air traffic management systems to effectively address the changing requirements associated with the integration of UAVs.

### *8.6.9 Establish Ethical and Privacy Guidelines*

Governmental bodies and industry groups must collaborate to establish and execute all-encompassing ethical principles and privacy legislation that are explicitly tailored to UAV activities [80]. The rules should comprehensively address data ownership, informed consent, retention, and secure data-handling methods. It is imperative to conduct privacy impact assessments for UAV missions that entail data collection in urban settings or near locations of heightened sensitivity [81]. Establishing a framework that facilitates responsible operations of UAVs while upholding individual privacy rights is crucial for cultivating public confidence and nurturing ethical conduct within the sector.

## 8.7 Conclusion

This study presented a performance evaluation model to analyze the most common technologies in UAV diversity combining in enhancing both Rayleigh and Ricean based on the diversity combination properties. In UAV communication, the Rayleigh and Ricean schemes have been used for a while in analyzing the communication diversities, however, on a smaller baseline. The study further offers a deep survey on the models and their related issues that challenge UAVs; the simulated results demonstrated that the Rayleigh model performs better when subjected to a bit error rate outstood the Ricean model. UAVs are evolving rapidly to emerge as indispensable weapons of war. Strategic technological areas must be identified early and built up as the technology matures to stay ahead in the future unmanned battlefield. Otherwise, UAV and unmanned technologies developments may outpace our capacity to assimilate them and formulate coherent and effective warfighting strategies. Weaponized UAVs have become a standard fixture of the defense forces worldwide. Stealth and swarm technologies are now coming into prominence. Governments are also looking at drones for homeland security applications such as disaster mitigation and border security. Some of these potential applications have sparked new debates about the scope of governmental surveillance, the possibility of abuse, and the independence of citizens.

## References


1. Shafik W, Matinkhah SM, Afolabi SS, Sanda MN (2020) A 3-dimensional fast machine learning algorithm for mobile unmanned aerial vehicle base stations. Int J Adv Appl Sci 2252(8814):8814
2. Marinho MA, Ferreira RSJ, da Costa JPC, de Freitas EP, Antreich F, Liu K, So HC, de Junior Sousa RT, Zelenovsky R (2013) Antenna array based positioning scheme for unmanned aerial vehicles. In: WSA 2013; 17th international ITG workshop on smart antennas, pp 1–6, VDE
3. Jangir D, Shankar G, Jain BB, Soni GK (2020) Performance analysis of lte system for 2x2 rayleigh and rician fading channel. In: 2020 international conference on smart electronics and communication (ICOSEC), pp 961–966, IEEE
4. Shafik W (2023) Cyber security perspectives in public spaces: drone case study. In: Handbook of research on cybersecurity risk in contemporary business systems, pp 79–97, IGI Global
5. Chen Z, Zeng M, Fei Z (2023) Joint unmanned aerial vehicle location and beamforming and caching optimization for cache-enabled multi-unmanned-aerial-vehicle networks. Electronics 12(16):3438
6. Hayawi K, Anwar Z, Malik AW, Trabelsi Z (2023) Airborne computing: a toolkit for uav-assisted federated computing for sustainable smart cities. IEEE Int Things J
7. Matinkhah SM, Shafik W (2019) Smart grid empowered by 5g technology. In: 2019 smart grid conference (SGC), pp 1–6, IEEE
8. Shafik W, Matinkhah SM, Ghasemazade M (2019) Fog-mobile edge performance evaluation and analysis on internet of things. J Adv Res Mobile Comput 1(3) 1–17
9. Shafik W, Mostafavi SA (2019) Knowledge engineering on internet of things through reinforcement learning. Int J Comput Appl 177(44):0975–8887
10. Shafik W, Matinkhah M, Asadi M, Ahmadi Z, Hadiyan Z (2020) A study on internet of things performance evaluation. J Commun Technol Electr Comput Sci 2020:1–19
11. Shafik W, Matinkhah SM, Ghasemzadeh M (2020) Theoretical understanding of deep learning in uav biomedical engineering technologies analysis. SN Comput Sci 1:1–13
12. Kaviri S, Tahsiri A, Taghirad HD (2021) A cooperative control framework of multiple unmanned aerial vehicles for dynamic oil spill cleanup. J Braz Soc Mech Sci Eng 43(6):289
13. Shafik W, Matinkhah SM (2019) Privacy issues in social web of things. In: 2019 5th international conference on web research (ICWR), pp 208–214, IEEE
14. Pliego-Jiménez J (2021) Quaternion-based adaptive control for trajectory tracking of quadrotor unmanned aerial vehicles. Int J Adapt Control Signal Process 35(5):628–641

15. Shafik W (2023) A comprehensive cybersecurity framework for present and future global information technology organizations. In: Effective cybersecurity operations for enterprise-wide systems, pp 56–79, IGI Global
16. Li K, Emami Y, Ni W, Tovar E, Han Z (2020) Onboard deep deterministic policy gradients for online flight resource allocation of uavs. IEEE Netw Lett 2(3):106–110
17. Yi M, Wang X, Liu J, Zhang Y, Bai B (2020) Deep reinforcement learning for fresh data collection in uav-assisted iot networks. In: IEEE INFOCOM 2020-IEEE conference on computer communications workshops (infocom WKSHPS), pp 716–721, IEEE
18. Wu F, Zhang H, Wu J, Song L (2020) Cellular uav-to-device communications: Trajectory design and mode selection by multi-agent deep reinforcement learning. IEEE Trans Commun 68(7):4175–4189
19. Shafik W, Matinkhah M, Etemadinejad P, Sanda MN (2020) Reinforcement learning rebirth, techniques, challenges, and resolutions. JOIV: Int J Inf Vis 4(3):127–135
20. Wang C, Wang J, Wang J, Zhang X (2020) Deep-reinforcement-learning-based autonomous uav navigation with sparse rewards. IEEE Internet Things J 7(7):6180–6190
21. Mukherjee M, Kumar V, Lat A, Guo M, Matam R, Lv Y (2020) Distributed deep learning-based task offloading for uav-enabled mobile edge computing. In: IEEE INFOCOM 2020-IEEE conference on computer communications workshops (INFOCOM WKSHPS), pp 1208–1212, IEEE
22. Liu CH, Piao C, Tang J (2020) Energy-efficient uav crowdsensing with multiple charging stations by deep learning. In: IEEE INFOCOM 2020-IEEE conference on computer communications, pp 199–208, IEEE
23. Das LB, Lijiya A, Jagadanand G, Aadith A, Gautham S, Mohan V, Reuben S, George G (2020) Human target search and detection using autonomous uav and deep learning. In: 2020 IEEE international conference on industry 4.0, artificial intelligence, and communications technology (IAICT), pp 55–61, IEEE
24. Zhao N, Cheng Y, Pei Y, Liang Y-C, Niyato D (2020) Deep reinforcement learning for trajectory design and power allocation in uav networks. In: ICC 2020-2020 IEEE international conference on communications (ICC), pp 1–6, IEEE
25. Tong P, Liu J, Wang X, Bai B, Dai H (2020) Deep reinforcement learning for efficient data collection in uav-aided internet of things. In: 2020 IEEE international conference on communications workshops (ICC Workshops), pp 1–6, IEEE
26. Liu CH, Ma X, Gao X, Tang J (2019) Distributed energy-efficient multi-uav navigation for long-term communication coverage by deep reinforcement learning. IEEE Trans Mob Comput 19(6):1274–1285
27. Li K, Ni W, Tovar E, Guizani M (2020) Deep reinforcement learning for real-time trajectory planning in uav networks. In: 2020 international wireless communications and mobile computing (IWCMC), pp 958–963, IEEE
28. Li K, Ni W, Tovar E, Jamalipour A (2020) Deep q-learning based resource management in uav-assisted wireless powered iot networks. In: ICC 2020-2020 IEEE international conference on communications (ICC), pp 1–6, IEEE
29. Guo W (2020) Partially explainable big data driven deep reinforcement learning for green 5g uav. In: ICC 2020-2020 IEEE international conference on communications (ICC), pp 1–7, IEEE
30. Chen D, Wei Y, Wang L, Hong CS, Wang L-C, Han Z (2020) Deep reinforcement learning based strategy for quadrotor uav pursuer and evader problem. In: 2020 IEEE international conference on communications workshops (ICC Workshops), pp 1–6, IEEE
31. Zhang Y, Zhuang Z, Gao F, Wang J, Han Z (2020) Multi-agent deep reinforcement learning for secure uav communications. In: 2020 IEEE wireless communications and networking conference (WCNC), pp 1–5, IEEE
32. Tetila EC, Machado BB, Menezes GK, Oliveira ADS, Alvarez M, Amorim WP, Belete NADS, Da Silva GG, Pistori H (2019) Automatic recognition of soybean leaf diseases using uav images and deep convolutional neural networks. IEEE Geosci Remote Sens Lett 17(5):903–907

33. Khelifi M, Butun I (2022) Swarm unmanned aerial vehicles (suavs): a comprehensive analysis of localization, recent aspects, and future trends. J Sensors 2022:1–10
34. Wan L, Liu R, Sun L, Nie H, Wang X (2022) Uav swarm based radar signal sorting via multi-source data fusion: a deep transfer learning framework. Inf Fusion 78:90–101
35. Ayass T, Coqueiro T, Carvalho T, Jailton J, Araújo J, Francês R (2022) Unmanned aerial vehicle with handover management fuzzy system for 5g networks: challenges and perspectives. Intell Robot 2(1):20–36
36. Shafik W, Matinkhah SM, Sanda MN, Shokoor F (2021) Internet of things-based energy efficiency optimization model in fog smart cities. JOIV: Int J Inf Vis 5(2):105–112
37. Butilă EV, Boboc RG (2022) Urban traffic monitoring and analysis using unmanned aerial vehicles (uavs): a systematic literature review. Remote Sens 14(3):620
38. Tan Y, Wang J, Liu J, Kato N (2022) Blockchain-assisted distributed and lightweight authentication service for industrial unmanned aerial vehicles. IEEE Internet Things J 9(18):16928–16940
39. Ahansal Y, Bouziani M, Yaagoubi R, Sebari I, Sebari K, Kenny L (2022) Towards smart irrigation: A literature review on the use of geospatial technologies and machine learning in the management of water resources in arboriculture. Agronomy 12(2):297
40. Faraji A, Rashidi M, Agha TMH, Rahnamayiezekavat P, Samali B (2022) Quality management framework for housing construction in a design-build project delivery system: a bim-uav approach. Buildings 12(5):554
41. Hu X, Li D (2020) Research on a single-tree point cloud segmentation method based on uav tilt photography and deep learning algorithm. IEEE J Sel Top Appl Earth Obser Remote Sens 13:4111–4120
42. Li X, Wang Q, Liu J, Zhang W (2020) Trajectory design and generalization for uav enabled networks: a deep reinforcement learning approach. In: 2020 IEEE wireless communications and networking conference (WCNC), pp 1–6, IEEE
43. Ghavimi F, Jantti R (2020) Energy-efficient uav communications with interference management: deep learning framework. In: 2020 IEEE wireless communications and networking conference workshops (WCNCW), pp 1–6, IEEE
44. Zhou S, Li B, Ding C, Lu L, Ding C (2020) An efficient deep reinforcement learning framework for uavs. In: 2020 21st international symposium on quality electronic design (ISQED), pp 323–328, IEEE
45. Walker O, Vanegas F, Gonzalez F, Koenig S (2020) Multi-uav target-finding in simulated indoor environments using deep reinforcement learning. In: 2020 IEEE aerospace conference, pp 1–9, IEEE
46. Sakir RKA, Ramli MR, Lee J-M, Kim D-S (2020) Uav-assisted real-time data processing using deep q-network for industrial internet of things. In: 2020 international conference on artificial intelligence in information and communication (ICAIIC), pp 208–211, IEEE
47. Huang H, Yang Y, Wang H, Ding Z, Sari H, Adachi F (2019) Deep reinforcement learning for uav navigation through massive mimo technique. IEEE Trans Veh Technol 69(1):1117–1121
48. Gao N, Qin Z, Jing X, Ni Q, Jin S (2019) Anti-intelligent uav jamming strategy via deep q-networks. IEEE Trans Commun 68(1):569–581
49. Zhao N, Liu Z, Cheng Y (2020) Multi-agent deep reinforcement learning for trajectory design and power allocation in multi-uav networks. IEEE Access 8:139670–139679
50. Zhang J, Yu Z, Mao S, Periaswamy SC, Patton J, Xia X (2020) Iadrl: Imitation augmented deep reinforcement learning enabled ugv-uav coalition for tasking in complex environments. IEEE Access 8:102335–102347
51. Yang Q, Zhang J, Shi G, Hu J, Wu Y (2019) Maneuver decision of uav in short-range air combat based on deep reinforcement learning. IEEE Access 8:363–378
52. Wu Q, Zhou Y, Wu X, Liang G, Ou Y, Sun T (2020) Real-time running detection system for uav imagery based on optical flow and deep convolutional networks. IET Intel Transport Syst 14(5):278–287
53. Tang J, Song J, Ou J, Luo J, Zhang X, Wong K-K (2020) Minimum throughput maximization for multi-uav enabled wpcn: a deep reinforcement learning method. IEEE Access 8:9124–9132

54. Qi H, Hu Z, Huang H, Wen X, Lu Z (2020) Energy efficient 3-d uav control for persistent communication service and fairness: a deep reinforcement learning approach. IEEE Access 8:53172–53184
55. Li B, Wu Y (2020) Path planning for uav ground target tracking via deep reinforcement learning. IEEE Access 8:29064–29074
56. Bah MD, Hafiane A, Canals R (2019) Crownet: Deep network for crop row detection in uav images. IEEE Access 8:5189–5200
57. Wang Q, Zhang W, Liu Y, Liu Y (2019) Multi-uav dynamic wireless networking with deep reinforcement learning. IEEE Commun Lett 23(12):2243–2246
58. Li K, Ni W, Tovar E, Jamalipour A (2019) On-board deep q-network for uav-assisted online power transfer and data collection. IEEE Trans Veh Technol 68(12):12215–12226
59. Kellenberger B, Marcos D, Lobry S, Tuia D (2019) Half a percent of labels is enough: Efficient animal detection in uav imagery using deep cnns and active learning. IEEE Trans Geosci Remote Sens 57(12):9524–9533
60. Zhang W, Song K, Rong X, Li Y (2018) Coarse-to-fine uav target tracking with deep reinforcement learning. IEEE Trans Autom Sci Eng 16(4):1522–1530
61. Koushik A, Hu F, Kumar S (2019) Deep q-learning-based node positioning for throughput-optimal communications in dynamic uav swarm network. IEEE Trans Cogn Commun Netw 5(3):554–566
62. Chen M, Saad W, Yin C (2019) Echo-liquid state deep learning for 360° content transmission and caching in wireless vr networks with cellular-connected uavs. IEEE Trans Commun 67(9):6386–6400
63. Challita U, Saad W, Bettstetter C (2019) Interference management for cellular-connected uavs: a deep reinforcement learning approach. IEEE Trans Wireless Commun 18(4):2125–2140
64. Wang C, Wang J, Shen Y, Zhang X (2019) Autonomous navigation of uavs in large-scale complex environments: a deep reinforcement learning approach. IEEE Trans Veh Technol 68(3):2124–2136
65. Ma X, Zhang Y, Zhang W, Zhou H, Yu H (2022) Sdwbf algorithm: a novel pedestrian detection algorithm in the aerial scene. Drones 6(3):76
66. Mitchell D, Blanche J, Harper S, Lim T, Gupta R, Zaki O, Tang W, Robu V, Watson S, Flynn D (2022) A review: challenges and opportunities for artificial intelligence and robotics in the offshore wind sector. Energy AI 8:100146
67. Won J, Kim D-Y, Park Y-I, Lee J-W (2022) A survey on uav placement and trajectory optimization in communication networks: from the perspective of air-to-ground channel models. ICT Express
68. Singh PK, Sharma A (2022) An intelligent wsn-uav-based iot framework for precision agriculture application. Comput Electr Eng 100:107912
69. Dang Y, Benzaid C, Yang B, Taleb T (2022) Deep learning for gps spoofing detection in cellular enabled unmanned aerial vehicle systems. arXiv:2201.00568
70. Liu Y, Han F, Zhao S (2022) Flexible and reliable multiuser swipt iot network enhanced by uav-mounted intelligent reflecting surface. IEEE Trans Reliab 71(2):1092–1103
71. Kaivosoja J (2022) Future possibilities and challenges for uav-based imaging development in smart farming. In: New developments and environmental applications of drones: proceedings of FinDrones 2020, pp 109–119, Springer
72. Kaivosoja J (2022) Future possibilities and challenges for uav-based imaging development in smart farming. In: New developments and environmental applications of drones: proceedings of FinDrones 2020, pp 109–119, Springer
73. Ameli Z, Aremanda Y, Friess WA, Landis EN (2022) Impact of uav hardware options on bridge inspection mission capabilities. Drones 6(3):64
74. Diaz Linares I, Pardo A, Patch E, Dehghantanha A, Choo K-KR (2022) Iot privacy, security and forensics challenges: an unmanned aerial vehicle (uav) case study. Handbook of big data analytics and forensics, pp 7–39
75. Abdel Hakeem SA, Hussein HH, Kim H (2022) Security requirements and challenges of 6g technologies and applications. Sensors 22(5):1969

76. Diaz Linares I, Pardo A, Patch E, Dehghantanha A, Choo K-KR (2022) Iot privacy, security and forensics challenges: an unmanned aerial vehicle (uav) case study. Handbook of big data analytics and forensics, pp 7–39
77. Wang X, Liu X, Wu J, Ju W, Chen X, Shen L (2022) Joint user scheduling, power configuration and trajectory planning strategy for uav-aided wsns. ACM Trans Sensor Netw 19(1):1–27
78. binti Burhanuddin LA, Liu X, Deng Y, Challita U, Zahemszky A (2022) Qoe optimization for live video streaming in uav-to-uav communications via deep reinforcement learning. IEEE Trans Veh Technol 71(5):5358–5370
79. Aslam MM, Du L, Ahmed Z, Azeem H, Ikram M (2019) Consensus performance of traffic management system for cognitive radio network: an agent control approach. In: Cyberspace data and intelligence, and cyber-living, syndrome, and health: international 2019 cyberspace congress, CyberDI and CyberLife, Beijing, China, December 16–18, 2019, Proceedings, Part II 3, pp 138–145, Springer
80. Aslam MM, Du L, Zhang X, Chen Y, Ahmed Z, Qureshi B (2021) Sixth generation (6g) cognitive radio network (crn) application, requirements, security issues, and key challenges. Wirel Commun Mob Comput 2021:1–18
81. Aslam MM, Zhang J, Qureshi B, Ahmed Z (2021) Beyond6g-consensus traffic management in crn, applications, architecture and key challenges. In: 2021 IEEE 11th international conference on electronics information and emergency communication (ICEIEC), pp 182–185, IEEE

# Chapter 9
# Intelligent Security for UAV-Assisted RIS Networks

**Ahmad Almadhor, Umair Ahmad Mughal, Ishtiaq Ahmad, Muhammad Ali Jamshed, and Ahmed Alkhayyat**

**Abstract** This chapter investigates methods for securing Unmanned Aerial Vehicle-assisted Reconfigurable Intelligent Surface (UAV-RIS) systems in next-generation communication networks. We propose a sophisticated deep reinforcement learning framework named Long Short-Term Memory Deep Deterministic Policy Gradient with Curiosity-Driven Exploration (LSTM-DDPG-CDE) to enhance security and ensure reliable communication within UAV-assisted RIS networks. By leveraging curiosity-driven intrinsic rewards, the framework encourages exploration of uncertain and novel network states, enabling the agent to detect and adapt to potential threats more effectively. The LSTM-DDPG-CDE architecture captures temporal dependencies in UAV trajectories, RIS phase configurations, and dynamic wireless channels, improving learning efficiency, convergence speed, and robustness against adversarial attacks. Through extensive simulations, we demonstrate that integrating UAV-RIS with the LSTM-DDPG-CDE framework significantly mitigates malicious

---

A. Almadhor
Department of Computer Engineering and Networks, College of Computer and Information Sciences, Jouf University, Sakakah, Saudi Arabia
e-mail: aaalmadhor@ju.edu.sa

U. A. Mughal (✉)
School of Computer Science and Information Systems, Northwest Missouri State University, Maryville, MO, USA
e-mail: umughal@nwmissouri.edu

I. Ahmad
Faculty of Electrical Engineering, Czech Technical University in Prague, Prague, Czech Republic
e-mail: ishtiaq001@gmail.com; ahmadish@fel.cvut.cz; ishtiaq.ahmad@seecs.edu.pk

School of Electrical Engineering and Computer Science (SEECS), National University of Science and Technology (NUST), Islamabad, Pakistan

M. A. Jamshed
College of Science and Engineering, University of Glasgow, Glasgow, UK
e-mail: muhammadali.jamshed@glasgow.ac.uk

A. Alkhayyat
Department of Computers Techniques Engineering, College of Technical Engineering, The Islamic University, Najaf, Iraq

© The Author(s), under exclusive license to Springer Nature Singapore Pte Ltd. 2026 
I. Ahmad et al. (eds.), *Reconfigurable Intelligent Surfaces for Unmanned Aerial Vehicle Communication*, Unmanned System Technologies,
https://doi.org/10.1007/978-981-96-9662-8_9

actions, such as signal nullification, phase manipulation, and eavesdropping, outperforming traditional reinforcement learning approaches.

**Keywords** UAVs · RIS · LSTM · DDPG · Curiosity-driven exploration

## 9.1 Introduction

The evolution of next-generation networks is set to transform mobile communications by integrating distributed network intelligence and enabling intricate spectrum coexistence across both passive and active radio services. Unmanned Aerial Vehicles (UAVs) possess unique advantages such as high maneuvrability, controllable trajectories, cost-efficient deployment and maintenance, and the capability to establish unobstructed line-of-sight (LoS) links. Such attributes position UAVs as pivotal components in 5G and future communication networks [1]. Numerous studies demonstrate that UAV-assisted wireless networks can substantially improve throughput, reliability, and security [2].

Concurrently, Reconfigurable Intelligent Surfaces (RIS) emerged as a transformative technology for wireless communication. RIS consists of programmable analogue elements capable of dynamically reflecting electromagnetic waves to control signal propagation and optimize channel gains. RIS are deployed on various surfaces, including building facades, vehicles, and indoor walls, with minimal deployment overhead. By carefully configuring these surfaces, RIS passively enhances the radio environment, improves communication links, and increases data rates. Integrating RIS with UAVs leverages both technologies to establish resilient and efficient communication links in complex network scenarios.

UAV-RIS networks are particularly promising for enhancing cybersecurity in diverse applications, including vehicular-to-everything (V2X) communications, public safety networks, and the Internet of Things (IoT). By dynamically controlling propagation conditions, RIS mitigates interference, improves signal quality, and strengthens reliability. Such capabilities are crucial in environments where densely connected devices are susceptible to security threats. UAVs act as mobile sensing and monitoring platforms within UAV-RIS systems. Equipped with sensors, cameras, and communication modules, UAVs detect, monitor, and respond to cyber threats in real time. UAV-RIS observes network variability, identifies potential intrusions, and facilitates adaptive security interventions, leveraging RIS to enhance network protection.

The synergy between UAVs and RIS in V2X, public safety, and IoT networks yields multiple security benefits. RIS elements enhance signal strength and quality, reducing the risk of unauthorized access or malicious interference. UAVs serve as agile security agents, identifying and responding to security violations promptly. Their mobility allows coverage of remote or isolated areas, extending monitoring capabilities and strengthening overall network security. Together, UAVs and RIS provide complementary strengths: RIS improves coverage and channel quality, while UAVs extend dynamic observation and intervention, enhancing threat detection and response capabilities.

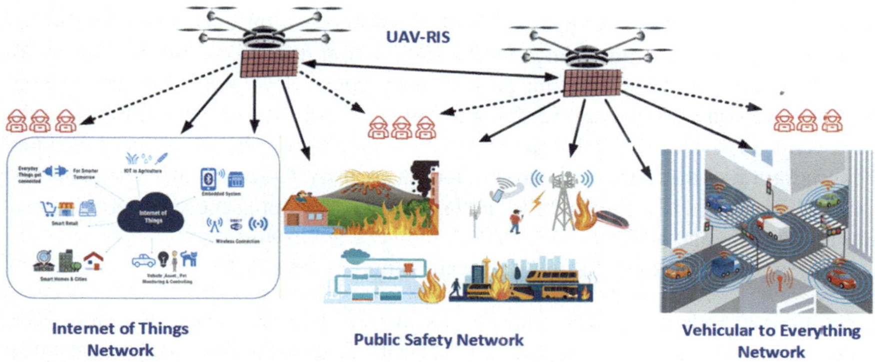


**Fig. 9.1** Intelligent security using UAV-RIS systems for IoT, public safety, and V2X networks

UAVs also adapt positions to focus on critical areas, optimizing coverage and surveillance efficiency. By prioritizing both security and reliability, UAV-RIS networks provide a robust platform for safe and efficient communication. The integration of UAVs with RIS enables improved network coverage, higher user access capacity, and enhanced physical layer security [2]. Previous works have explored various resource allocation strategies in next-generation networks [3–13]. However, these advancements also introduce new vulnerabilities, as adversaries exploit the programmability of RIS and the mobility of UAVs to launch sophisticated attacks. Figure 9.1 illustrates a typical operational scenario of a UAV-RIS system, highlighting the dual nature of RIS programmability, which can both enhance and threaten network security. Previous works have explored various resource allocation and optimization strategies in next-generation wireless networks, including RIS-assisted mmWave systems, multi-user MIMO, and ISAC-enabled communication [14–18]. These studies demonstrated significant gains in spectral efficiency and adaptive beamforming performance, yet they did not explicitly address dynamic security threats or adversarial conditions in UAV-RIS networks.

Motivated by the increasing security challenges in dynamic UAV-RIS communication environments, this chapter proposes a robust deep reinforcement learning framework based on a Long Short-Term Memory Deep Deterministic Policy Gradient architecture enhanced with Curiosity-Driven Exploration (LSTM-DDPG-CDE). In our previous work [28], we employed an LSTM-DDPG framework to capture temporal correlations in UAV-RIS network states and enable adaptive decision-making in dynamic wireless environments. While the framework demonstrated promising performance, its exploration strategy relied primarily on conventional noise-based methods, which can lead to inefficient exploration and slower convergence in complex and high-dimensional state spaces. Moreover, the earlier approach did not explicitly encourage exploration of uncertain or rarely visited network states, potentially limiting the agent's ability to adapt to stealthy attacks or rapidly changing network conditions.

This chapter builds upon and significantly extends our prior work by incorporating a curiosity-driven exploration mechanism that provides intrinsic rewards for visiting novel or uncertain states. This intrinsic motivation encourages the learning agent to actively explore informative state-action transitions, improving exploration efficiency and accelerating policy convergence. By jointly leveraging temporal feature learning through LSTM and intrinsic curiosity-driven exploration, the proposed LSTM-DDPG-CDE framework dynamically adapts UAV trajectories and RIS configurations to enhance communication security. Consequently, the proposed approach improves learning stability and robustness, enabling more effective operation in dynamic and adversarial UAV-RIS network environments.

The key contributions of this chapter are summarized as follows:

- We analyze potential adversarial threats in UAV-RIS networks, including RIS-assisted attacks, and identify emerging security challenges that could compromise communication reliability and confidentiality.
- We leverage the LSTM-DDPG framework to capture temporal correlations in UAV-RIS network states. The actor and critic networks are carefully designed to model time-dependent dynamics in UAV trajectories, RIS phase shifts, channel variations, and overall network behavior.
- Unlike prior approaches relying on experience replay, our framework incorporates CDE, which assigns intrinsic rewards for visiting uncertain or novel states. This encourages the agent to explore rare or stealthy attack scenarios, enabling efficient learning of robust security policies in dynamic environments.
- Extensive simulations demonstrate that integrating UAV-RIS with the proposed LSTM-DDPG-CDE framework effectively mitigates malicious attacks, including signal nullification, RIS phase manipulation, and eavesdropping, achieving superior security performance compared to traditional reinforcement learning methods.

## 9.2 Utilizing RIS for UAV Communications: Exploring Applications and Security Ramifications

### *9.2.1 Expanding Coverage via UAV-RIS*

UAV-RIS offers the potential to broaden coverage and improve spectral efficiency within cellular networks. By enabling precise control over signal directionality, RIS can effectively expand range of base stations into previously inaccessible areas, ensuring comprehensive 360-degree zone. However the configuration is vulnerable to bluffing attacks [29]. An adversary could exploit a reprobate UAV-supported RIS system to manipulate transmission pathways, leading to potential network disturbances or the interception of confidential information. Recent research has also investigated AI-driven anomaly detection, intrusion prevention, and quantum machine learning for IoT and vehicular networks [19–23]. These works showed the potential of intelligent frameworks to enhance robustness and detect stealthy attacks; however, most approaches rely on conventional exploration strategies and do not fully leverage temporal correlations in network states.

Furthermore, several studies focused on task offloading, resource allocation, and covert/secure communication in UAV, edge, and RIS-aided networks [24–27]. While these contributions improve adaptability and resource efficiency, they do not incorporate intrinsic curiosity-driven mechanisms for proactive exploration under uncertainty. However, in the proposed framework (LSTM-DDPG-CDE), which allows the agent to efficiently explore underrepresented or uncertain UAV-RIS network states, thereby learning robust security and resource allocation policies while adapting to dynamic and adversarial environments.

### *9.2.2 Enhanced Capacity with UAV-RIS*

Utilizing UAV-supported RIS systems can substantially enhance network throughput and capacity. Through the manipulation of phase shifts among RIS elements, the system can effectively mitigate interference, contingent upon the availability of accurate Channel State Information (CSI) at the RIS controller. However introduces a susceptibility to potential jamming attacks. UAV-supported RIS offers a solution to the substantial access demands arising from the proliferation of IoT devices. Through the manipulation of indoor wireless channels to prioritize certain users, effectively bolster system capacity [30]. Nevertheless, remain susceptible to Denial-of-Service (DoS) attacks, wherein an adversary might inundate the network with spurious access requests, leading to disruptions in service.

RIS technology has the potential to facilitate effective spectrum sharing through interference reduction. Nevertheless, this functionality also opens doors for malicious actors to carry out spoofing assaults. By tampering with the RIS phase shifters, an attacker could disorder the network's band distribution, leading to interference plus detrimental impacts on system functionality. Utilizing UAV-supported RISs enhances the performance of land-based cellular networks by fostering strong Line-of-Sight (LoS) connections between airborne and terrestrial units. Nevertheless, this LoS feature could be utilized by malicious parties for data interception purposes. Moreover, the RIS might be manipulated to generate harmful reflected signals, reducing the Signal-to-Noise Ratio (SNR) at targeted areas and minimizing interception opportunities. Even with these precautions, determined and sophisticated adversaries may discover vulnerabilities within these systems, emphasizing the imperative for resilient security measures.

## 9.3 Potential Weaknesses and Threats in UAV-Enhanced RIS-Assisted Systems

RIS and UAV-assisted wireless networks are central to emerging 6G and Integrated Sensing and Communication (ISAC) architectures. While these technologies promise unprecedented spectral efficiency, coverage enhancement, and sensing capabilities, they also introduce novel attack surfaces fundamentally different from conventional wireless systems. The combination of a dynamic RIS phase control and mobile UAV deployment creates vulnerabilities across the physical layer, control plane, and machine-learning-enabled network intelligence.

### 9.3.1 Advanced Phase and Reflection Manipulation

RIS elements dynamically adjust the phase shifts to manipulate electromagnetic wave propagation [36–39]. Malicious manipulation of these phase shifts systematically distorts legitimate communication channels. Unlike traditional jamming, these attacks do not require high transmission power but rely on precise phase and amplitude alterations [39]. For instance, selective nulling produces destructive interference at a target receiver, creating deep fades without raising interference alarms. Spatially precise attacks concentrate interference in specific directions or zones, which conventional jammers cannot replicate. UAV mobility amplifies these threats, allowing adversaries to reposition either themselves or RIS units to maximize destructive impact while remaining covert [40, 41].

### 9.3.2 Eavesdropping and Channel Exploitation

RIS-assisted networks present unique risks for signal interception. Carefully tuned RIS phases redirect sidelobes or reflected signals toward unintended receivers, enabling covert eavesdropping. When multiple adversarial receivers collude, they reconstruct channel behavior and infer confidential information, particularly if RIS phase updates are predictable or poorly secured [34]. The mobility of UAVs allows attackers to adapt trajectories in real time, optimizing interception paths and maintaining line-of-sight to RIS-reflected signals, further enhancing eavesdropping efficiency [42, 43].

### 9.3.3 Control-Plane and Signaling Vulnerabilities

The UAV-RIS control link is a critical security point. Compromising the control interface allows adversaries to inject incorrect configuration commands, alter beamforming vectors, or disrupt RIS updates [44, 45]. Protocol manipulation attacks targeting signaling channels force UAVs to perform unintended maneuvers or misconfigure RIS phases. Distributed denial-of-service (DDoS) attacks on control channels stall real-time updates, severely degrading system performance. Emerging research emphasizes the need for quantum-resilient cryptography in UAV-RIS signaling to protect against both current and future quantum-enhanced attacks [46, 47].

### 9.3.4 Machine Learning and Adversarial AI Threats

Deep learning and reinforcement learning algorithms, often used for UAV trajectory planning and RIS phase configuration, introduce additional attack surfaces [31]. Deep neural networks (DNNs) are deceived with adversarial inputs, leading to misalignment of beams or poor UAV navigation [48–50]. Poisoning of training data, such as manipulated channel state information, embeds persistent vulnerabilities into deployed models. Deep reinforcement learning (DRL) agents, such as DDPG or A2C-based UAV controllers, are particularly susceptible to reward manipulation or state perturbation, which mislead the agent into trajectories that compromise coverage or expose communications to interception [51, 52].

### 9.3.5 Physical Layer Authentication and Covert Communication Risks

RIS-assisted systems frequently rely on physical layer authentication (PLA) and covert communication mechanisms. Adversarial RIS manipulation induces rapid fluctuations in received signal strength, phase, or SNR, undermining PLA and enabling identity spoofing [53, 54]. Advanced RIS configurations, including curved or active RIS, can create stealthy communication channels that are nearly invisible to standard monitoring systems. Active RIS can also unintentionally amplify malicious signals or noise, magnifying disruptive effects when under adversarial control [55, 56].

### 9.3.6 RIS-Assisted ISAC Vulnerabilities

Integration of RIS into ISAC systems introduces novel risks. Since RIS states influence both communication and sensing, malicious manipulation generates false environmental feedback, misleading UAV radar or positioning subsystems [31]. In autonomous UAV swarms, a single compromised UAV or RIS node can propagate erroneous information, disrupting swarm coordination and degrading collaborative sensing and communication tasks [32].

### 9.3.7 Swarm and Collaborative UAV Vulnerabilities

Multi-UAV swarms magnify potential attack consequences. Coordinated attacks targeting one UAV can cascade through the swarm, introducing false telemetry, misaligned beamforming vectors, or incorrect trajectory adjustments [33]. Such attacks not only degrade system performance but also cause mission-critical failures in applications such as surveillance, industrial IoT monitoring, or disaster response [57–59].

### 9.3.8 Security Implications and Mitigation Strategies

Mitigating these emerging threats requires a multi-layered approach. Robust physical layer security techniques, including artificial noise design and secure beamforming, counteract RIS-enabled interference. Quantum-resilient cryptographic protocols protect control-plane communications [60, 61]. AI-driven anomaly detection and intrusion detection systems tailored to RIS and UAV behavior are critical for identifying unusual patterns in phase shifts, trajectories, or channel characteristics. Secure and robust reinforcement learning frameworks ensure UAV path planning and RIS configuration remain resilient to adversarial manipulation while maintaining real-time adaptability and operational efficiency [62, 63]. As UAV-RIS networks evolve toward large-scale 6G deployments, continuous monitoring, adaptive defense mechanisms, and resilient design principles are essential to safeguard system reliability and security [64].

## 9.4 Proposed Framework for Intrusion Detection System Integrating LSTM-DDPG-CDE

This section introduces the proposed LSTM-DDPG-CDE framework for intrusion detection in UAV-assisted RIS networks. The primary goal of the framework is to enhance signal quality for legitimate users while effectively detecting and mitigating malicious activities from unauthorized actors. By integrating DDPG with LSTM networks and CDE, the framework captures temporal dynamics in the UAV-RIS environment and encourages exploration of novel or uncertain states, thereby improving security and reliability.

The proposed LSTM-DDPG-CDE procedure comprises several key components, including actor and critic networks, each incorporating an LSTM architecture. Both networks consist of online and target versions. The actor-network utilizes policy gradient methods to derive deterministic actions based on environmental observations, while the critic network evaluates these actions using a Q-function. Historical transition data is collected and used to compute an intrinsic curiosity reward, which guides the agent toward exploring rare or previously unseen network states, complementing the extrinsic reward for maintaining secure communications.

### 9.4.1 System Architecture

In UAV-assisted RIS-enabled communication systems, maintaining secure and reliable transmission is challenging due to the dynamic wireless environment, UAV mobility, and the presence of malicious actors attempting to compromise network integrity. We consider a communication network consisting of a set of users

$\mathcal{U} = \{u_1, u_2, \ldots, u_U\}$, a UAV equipped with an RIS $\mathcal{R}$, and a central BS. The UAV-RIS enhances signal quality for legitimate users while mitigating interference or eavesdropping from malicious nodes $\mathcal{M} = \{m_1, m_2, \ldots, m_M\}$. The received signal at the user $u_i$ is given as

$$y_i = \mathbf{h}_{\mathrm{UAV},u_i}^H \Theta \mathbf{h}_{\mathrm{BS,UAV}} x + \sum_{j=1}^{M} \mathbf{g}_{m_j,u_i}^H x_{m_j} + n_i, \tag{9.1}$$

where $x$ is the transmitted symbol, $\mathbf{h}_{\mathrm{BS,UAV}}$ is the channel between BS and UAV, $\Theta$ is the RIS phase shift matrix, $\mathbf{h}_{\mathrm{UAV},u_i}$ is the UAV-to-user channel, $\mathbf{g}_{m_j,u_i}$ is the malicious user channel, and $n_i$ is AWGN. The system has two objectives: the first is to maximize the secrecy rate, which is defined as

$$R_s^{(i)} = \left[C(\mathbf{h}_{\mathrm{UAV},u_i}^H \Theta \mathbf{h}_{\mathrm{BS,UAV}}) - \max_{m_j \in \mathcal{M}} C(\mathbf{g}_{m_j,u_i}^H)\right]^+. \tag{9.2}$$

The second objective of this chapter is to mitigate malicious activities, such as signal nullification, noise injection, RIS phase manipulation, or UAV trajectory exploitation. The system adapts UAV trajectories and RIS configurations to maintain secure communications.

### 9.4.2 *Problem Formulation*

The UAV-RIS system is modeled as a Markov Decision Process (MDP) $\langle \mathcal{S}, \mathcal{A}, \mathcal{P}, \mathcal{R} \rangle$, where $\mathcal{S}$ includes UAV location, RIS phase shifts, CSI of users and attackers, and historical signals. $\mathcal{A}$ consists of UAV trajectory adjustments, RIS phase updates, and power allocation. $\mathcal{P} : \mathcal{S} \times \mathcal{A} \times \mathcal{S} \to [0, 1]$ defines the state transition probability and $\mathcal{R} : \mathcal{S} \times \mathcal{A} \to \mathbb{R}$ is the reward:

$$r_t = \sum_{i=1}^{U} R_s^{(i)} - \lambda \sum_{j=1}^{M} \mathbb{I}\{m_j\} + \eta r_t^{\mathrm{CDE}}, \tag{9.3}$$

where $r_t^{\mathrm{CDE}}$ is the curiosity-driven intrinsic reward, $\lambda$ and $\eta$ are the weighting factors and $\mathbb{I}\{\cdot\}$ is the indicator function. The optimal policy $\pi^*$ maximizes cumulative rewards is defined as

$$\pi^* = \arg\max_{\pi} \mathbb{E}\left[\sum_{t=0}^{T} \gamma^t r_t \,\middle|\, \pi\right], \tag{9.4}$$

with discount factor $\gamma \in [0, 1]$ and time horizon $T$. The high-dimensional continuous action space and temporal dependencies make learning challenging, motivating the integration of LSTM for temporal modeling, DDPG for continuous control, and CDE for exploration of critical or rare attack scenarios (Fig. 9.2).

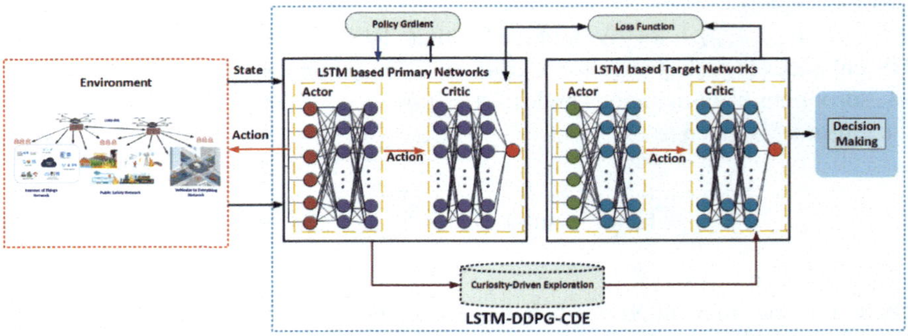


**Fig. 9.2** Proposed LSTM-DDPG-CDE framework for secured UAV-RIS system

### *9.4.3 Enhancing Temporal Dynamics in Actor-Critic Networks with LSTM-DDPG-CDE*

The actor-critic architecture employs separate LSTM-based actor and critic networks to process sequences of state observations $\{s_{t-k}, \ldots, s_t\}$. The hidden representation $h_t$ at time $t$ is defined as

$$h_t = \text{LSTM}(s_t, h_{t-1}). \tag{9.5}$$

Such a hidden representation $h_t$ captures long-term temporal dependencies of the UAV-RIS-assisted system. In the actor network, $h_t$ is mapped to deterministic actions, which are defined as

$$a_t = \mu(h_t|\theta^{\mu}), \tag{9.6}$$

and the critic evaluates $(h_t, a_t)$ pairs:

$$Q(h_t, a_t|\theta^{Q}), \tag{9.7}$$

where $\theta^{\mu}$ and $\theta^{Q}$ are network parameters. CDE introduces intrinsic rewards based on state prediction errors, guiding the agent toward novel or uncertain states. This mechanism enables the detection of stealthy attacks that may otherwise be overlooked. The final action fuses trajectory control, RIS phase adjustment, and power allocation through a modular operator $\oplus$, ensuring flexible yet stable control. The proposed LSTM-DDPG-CDE framework delivers three main advantages: (i) temporal dependency modeling for dynamic UAV-RIS environments, (ii) curiosity-driven exploration for rare or stealthy threats, and (iii) robust policy learning for secure communication in adversarial, time-varying wireless environments.

## 9.5 Performance Evaluation

In this section, we first describe the UAV-assisted RIS communication environment configuration and then present simulations to evaluate the effectiveness of the proposed framework across various attack scenarios.

### *9.5.1 Setting Up the UAV-Assisted RIS Environment*

Our primary objective is to develop strategies to mitigate diverse attacks, including signal nullification, noise injection, UAV-RIS link exploitation, and coordinated UAV swarm attacks. To achieve this, we employ deep reinforcement learning techniques for intrusion detection and threat mitigation. The methodology begins by establishing a simulated environment that replicates potential attacks, generating datasets containing both normal and compromised instances. These datasets serve as the foundation for training the LSTM-DDPG-CDE model.

The simulation incorporates UAV-assisted RIS communication, where a user moves along a predetermined path at constant velocity while encountering potential eavesdroppers. A UAV equipped with RIS relays information from a central station to the receiver. Central stations are positioned at fixed locations, while the UAV-RIS dynamically adjusts its reflective elements to optimize the communication link at a height of 20 m. Both LoS and non-LoS paths are considered, including multipath reflections. RIS plays a critical role by redirecting signals toward legitimate users, thereby enhancing reliability and enabling adaptive countermeasures against malicious attacks.

The secrecy rate is used to measure system performance under eavesdropping conditions, capturing both cyber and physical dimensions. Specifically, it represents the difference between the data rate received by the authorized user and the maximum data rate potentially accessible to eavesdroppers. The UAV-RIS framework thus jointly considers physical channel characteristics and cyber vulnerabilities to assess communication security.

### *9.5.2 Simulation Results*

We conducted extensive experiments to evaluate the proposed LSTM-DDPG-CDE framework in terms of intrusion detection and secure communication. The simulations are designed to address malicious threats and demonstrate the resilience of the proposed methodology.

Figure 9.3a shows the average secrecy rate as a function of the number of users. The proposed LSTM-DDPG-CDE framework consistently outperforms other machine learning models, including RL, DRL, DDPG, and LSTM-DDPG. The results high-

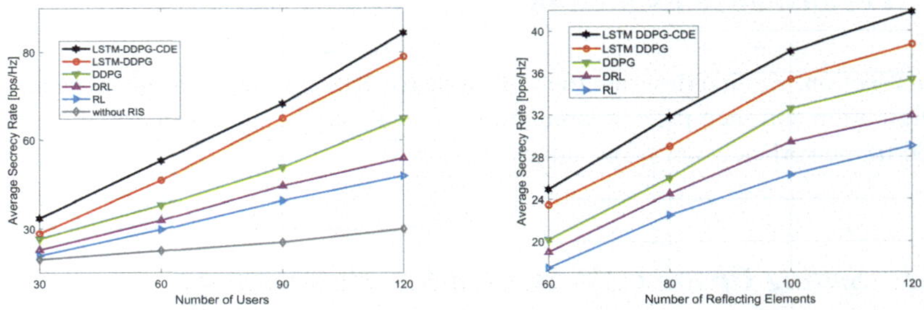


(a) Average secrecy rate versus number of users.

(b) Average secrecy rate versus number of reflecting elements.

**Fig. 9.3** Performance evaluation of the proposed LSTM-DDPG-CDE framework under varying network conditions

**Table 9.1** Comparison of intrusion detection performance under various malicious threats

| Models | Accuracy (%) | Precision (%) | Recall (%) | F1 (%) | AUC (%) |
|---|---|---|---|---|---|
| RL | 96.11 | 96.21 | 95.22 | 95.01 | 96.22 |
| DRL | 97.82 | 97.32 | 97.26 | 97.10 | 97.12 |
| DDPG | 98.61 | 97.55 | 97.58 | 97.75 | 98.63 |
| LSTM-DDPG | 99.01 | 98.11 | 98.31 | 98.01 | 99.01 |
| LSTM-DDPG-CDE | 99.52 | 99.04 | 99.08 | 98.75 | 99.53 |

light the effectiveness of our framework in maintaining secure communications even as network size grows, achieving higher secrecy rates by dynamically adapting UAV trajectories and RIS configurations to mitigate attacks.

Figure 9.3b illustrates the effect of the number of RIS reflecting elements on the average secrecy rate. As the number of elements increases, the proposed LSTM-DDPG-CDE framework achieves substantial performance gains compared to other methods. The results emphasize the importance of adaptive RIS control in securing UAV-assisted networks against sophisticated attacks.

Table 9.1 compares intrusion detection performance across several models using standard metrics such as accuracy, precision, recall, F1 score, and AUC. The proposed LSTM-DDPG-CDE model consistently achieves the highest performance across all metrics, confirming its superiority in detecting and mitigating malicious activities in dynamic UAV-RIS networks.

## 9.6 Conclusion

In this chapter, we highlight potential vulnerabilities in UAV-assisted RIS networks that can be exploited by malicious actors. To address these security challenges, we proposed the LSTM-DDPG-CDE framework, a deep reinforcement learning-based approach capable of dynamically detecting and mitigating cyber threats while optimizing UAV trajectories and RIS configurations. Extensive simulations demonstrate that the proposed framework effectively strengthens network security, enhances reliability, and ensures resilient communication in the presence of sophisticated attacks. The results confirm that integrating temporal modeling with continuous control and adaptive exploration through CDE significantly improves the system's capability to counteract malicious behaviors.

## References

1. Li S, Duo B, Yuan X, Liang Y-C, Di Renzo M (2020) Reconfigurable intelligent surface assisted unmanned aerial vehicles communication: joint trajectory design and passive beamforming. IEEE Wireless Commun Lett 9(5):716–720
2. Xu Y, Xie H, Wu Q, Huang C, Yuen C (2022) Robust max-min energy efficiency for reconfigurable intelligent surface-aided HetNets With distortion noises. IEEE Trans Commun 70(2):1457–1471
3. Ishtiaq I, Narmeen R, Aftab MW, Alkhrijah Y, Alawad MA, Kaushik A (2024) Random access optimization with generative adversarial networks in industrial IoT using deep deterministic policy gradient approach. IEEE Open J Soc Commun
4. Ishtiaq I, Narmeen R, Alawad MA, Alkhrijah Y, Ho P-H (2024) Integrating visual geometry and mask region CNN for enhanced UAV detection and identification. In: IEEE vehicular technology conference (VTC2024-Fall). Washington DC USA 2024. pp 1–6
5. Narmeen R, Mach P, Becvar Z, Ishtiaq I (2024) Joint exit selection and offloading decision for applications based on deep neural networks. IEEE Internet Things J 11(23):38098–38112
6. Ishtiaq I et al (2024) Optimizing cell association and stability in integrated aerial-to-ground next-generation consumer wireless networks. IEEE Trans Cons Electr 70(3):6262–6276
7. Ishtiaq I, Becvar Z, Mach P, Gesbert D (2024) Coordinated machine learning for energy efficient D2D communication. IEEE Wireless Commun Lett 13(5):1493–1497
8. Ishtiaq I et al (2024) Machine-learning-based optimal cooperating node selection for internet of underwater things. IEEE Internet Things J 11(12):22471–22482
9. Ishtiaq I, Becvar Z, Mach P (2024) Coordinated machine learning for channel reuse and transmission power allocation for D2D communication. In: IEEE global communications conference (Globecom)
10. Ishtiaq I, Narmeen R et al (2024) DRL-based resource management for task-centered semantic communication. In: IEEE international symposium on personal, indoor and mobile radio communications (PIMRC Workshop)
11. Ahmad I, Kaleem Z, Plachy J, Tabassum H, Becvar Z, Saad W (2024) Guest editorial: flexible reconfigurable intelligent surfaces for integrated air-to-ground internet of things: architecture, recent advances, and future applications. IEEE Internet Things Mag 7(2):12–13
12. Lyu B, Hoang DT, Gong S, Niyato D, Kim DI (2020) IRS-based wireless jamming attacks: when jammers can attack without power. IEEE Wireless Commun Lett 9(10):1663–1667
13. Huang C, Zappone A, Alexandropoulos GC, Debbah M, Yuen C (2019) Reconfigurable intelligent surfaces for energy efficiency in wireless communication. IEEE Trans Wireless Commun 18(8):4157–4170

14. Ahmad I et al (2026) Intelligent anomaly-aware adaptive post-quantum cryptography in V2X systems. In: Transactions on Consumer Electronics. Early Access. https://doi.org/10.1109/TCE.2026.3671160.
15. Du J, Ouyang Q, Li X, Ahmad I (2026) Unified tensor framework for RIS-aided mmWave systems: low-complexity channel estimation and low-rank feedback. IEEE Internet Things J, Accepted
16. Yan J, Zhang Y, Li X, Ahmad I (2026) Joint optimization for active RIS-assisted hybrid near- and far-field multi-user MIMO systems. IEEE Trans Veh Technol, Accepted
17. Li J, Yang L, Ahmad I et al (2026) Robust secure beam-scanning for near-field ISAC enabled by location division multiple access. Early Access, IEEE J Sel Areas Commun. https://doi.org/10.1109/JSAC.2026.3669090
18. Ahmad I, Becvar Z, Mach P (2026) Coordinated federated learning for radio resource management in future wireless networks. IEEE Wireless Commun Lett 15:1420–1424. https://doi.org/10.1109/LWC.2026.3654242
19. Ahmad I et al (2025) Quantum machine learning for anomaly detection: the future of smarter, safer IoT networks. IEEE Netw Early Access. https://doi.org/10.1109/MNET.2025.3637512
20. Sana MS, Ahmad I et al (2025) Secure UAV-RIS-enabled IoT systems: federated DDPG with attention mechanism for adversarial attack mitigation. Early Access, IEEE Internet Things J. https://doi.org/10.1109/JIOT.2025.3580969
21. Ahmad I, Wen M, Ho P-H, Yuen C (2025) Guest editorial of the special section on artificial intelligence enabled Internet of vehicular communications. IEEE Trans Consum Electron 71(4):10917–10920. https://doi.org/10.1109/TCE.2025.3598185
22. Ahmad I, Narmeen R, Tabassum H, Becvar Z (2025) Intelligent localization-based DDPG for UAV-borne RIS in vehicular communication. Early Access, IEEE Trans Veh Technol. https://doi.org/10.1109/TVT.2025.3578644
23. Ahmad I et al (2025) Quantum CNN for detection and identification of UAV-enabled non-terrestrial networks. IEEE Wirel Commun 32(3):28–36. https://doi.org/10.1109/MWC.001.2400419
24. Wu S, Yang L, Ahmad I et al (2025) DRL-based pricing-driven task offloading and dynamic resource allocation in vehicle edge computing. Early Access, IEEE Trans Mobile Comput. https://doi.org/10.1109/TMC.2025.3569817
25. Liang Y, Yang L, Ahmad I, Valkama M (2025) Covert transmission and physical-layer security of STAR-RIS-assisted uplink SGF-NOMA systems. Early Access, IEEE Trans Commun. https://doi.org/10.1109/TCOMM.2025.3565500
26. Ahmad I (2025) Intelligent task offloading decisions for enhanced swarm-based UAV surveillance. IEEE Trans Aerospace Electr Syst. Early Access (2025). https://doi.org/10.1109/TAES.2025.3564282
27. Li J, Yang L, Ahmad I et al (2025) Multi-layer transmitting RIS-aided receiver for collaborative jamming and anti-jamming networks. Early Access, IEEE Trans Wireless Commun. https://doi.org/10.1109/TWC.2025.3554386
28. Mughal UA, Alkhrijah Y, Almadhor A, Yuen C (2024) Deep learning for secure UAV-assisted RIS communication networks. IEEE Internet Things Mag 7(2):38–44
29. Zhi K, Pan C, Ren H, Chai KK, Elkashlan M (2022) Active reconfigurable intelligent surface versus passive reconfigurable intelligent surface: Which is superior with the same power budget? IEEE Commun Lett 26(5):1150–1154
30. Pan Y, Hou Y, Li M, Gerdes RM, Zeng K, Towfiq MA, Cetiner BA (2017) Message integrity protection over wireless channel: countering signal cancellation via channel randomization. IEEE Trans Dependable Secure Comput 17(1):106–120
31. Wang Y, Lu H, Zhao D, Deng Y, Nallanathan A (2022) Wireless communication in the presence of illegal reconfigurable intelligent surface: signal leakage and interference attack. IEEE Wireless Commun 29(3):131–138
32. Di Renzo M, Zappone A, Debbah M, Alouini M-S, Yuen C, De Rosny J, Tretyakov S (2020) Smart radio environments empowered by reconfigurable intelligent surfaces: how it works, state of research, and the road ahead. IEEE J Sel Areas Commun 38(11):2450–2525

33. Nie L, Ning Z, Wang X, Hu X, Cheng J, Li Y (2020) Data-driven intrusion detection for intelligent internet of vehicles: a deep convolutional neural network-based method. IEEE Trans Netw Sci Eng 7(4):2219–2230
34. Hu S, Chen X, Ni W, Hossain E, Wang X (2021) Distributed machine learning for wireless communication networks: techniques, architectures, and applications IEEE Commun Surv & Tutor 23(3):1458–1493
35. Hu S, Yuan X, Ni W, Wang X (2022) Trajectory planning of cellular-connected UAV for communication-assisted radar sensing. IEEE Trans Commun 70(9):6385–6396
36. Ahmad I et al (2023) Co-channel interference management for heterogeneous networks using deep learning approach. Information 14(139)
37. Narmeen R et al (2023) Co-channel interference evaluation for coexisting public-safety networks and 5G commercial network. J Smart Internet Things **2**
38. Narmeen R et al (2021) Shortest propagation delay-based relay selection for underwater acoustic sensor networks. IEEE Access 37923–37935
39. Ahmad U et al (2020) Cooperative resource management for C-V2I communications in a dense urban environment. Veh Commun
40. Ahmad I, Chang KH (2019) Effective SNR mapping and link adaptation strategy for next-generation underwater acoustic communications networks: a cross-layer approach. IEEE Access 7:44150–44164
41. Ahmad I, Chen W, Chang KH (2017) LTE-railway user priority-based cooperative resource allocations schemes for coexisting public safety and railway networks. IEEE Access 5:7958–8000
42. Ahmad I, Chang KH (2019) Downlink power allocation strategy for next-generation underwater acoustic communications networks. Electronics
43. Chen W et al (2017) Co-channel interference management using eICIC/FeICIC with coordinated scheduling for the coexistence of PS-LTE and LTE-R networks. EURASIP J Wireless Commun Netw 34:1–14
44. Ahmad I, Chang KH (2019) Mission critical user priority-based cooperative resource allocation schemes for multi-layer next generation public safety networks. Phys Commun
45. Ahmad I et al (2019) Quality-of-service aware game theory-based uplink power control for 5G heterogeneous networks. Mobile Netw Appl 24(2):556–563
46. He J et al (2019) Compressive sensing based random access for machine type communications considering tradeoff between link performance and latency. EURASIP J Wireless Commun Netw 1–11
47. Shi L et al (2018) Hidden Markov model-based drone sound recognition using MFCC technique in practical noisy environments. J Commun Netw 20(5):509–518
48. He Y et al (2019) SVM-based drone sound recognition using the combination of HLA and WPT techniques in practical noisy environment. KSII Trans Internet Inf Syst
49. Guo J et al (2020) Classification, positioning, and tracking of drones by HMM using acoustic circular microphone array beamforming. EURASIP J Wireless Commun Netw 9:1–19
50. Hu T et al (2020) 3D optimal surveillance trajectory planning for multi-UAVs using particle swarm optimization with surveillance area priority. IEEE Access
51. Ahmad I et al (2020) QoS priority-based mobile personal cell deployment with load balancing for interference reduction between users on coexisting public safety and railway LTE networks. Electronics 9(12):1–13
52. Chang KH et al (2020) Mission-critical user priority-based random access method and apparatus for collision resolution in coexisting PS-LTE and LTE-M network. US Patent US10645734B2
53. Ahmad I et al (2024) Integrating visual geometry and mask region CNN for enhanced UAV detection and identification. In: Proceeding of the IEEE vehicular technology conference (VTC)
54. Ahmad I et al (2024) DRL-based resource management for task-centered semantic communication. In: Proceeding of the IEEE international symposium on personal, indoor and mobile radio communications (PIMRC)

55. Ahmad I, Chang KH (2016) Design of system-level simulator architecture for underwater acoustic communications and networking. In: Proceeding of the IEEE international conference on information and communication technology convergence (ICTC), Jeju, Korea
56. Ahmad I, Chang KH (2017) Analysis on MIMO transmit diversity and multiplexing techniques for ship ad-hoc networks under a maritime channel model in coastline areas. In: Proceeding of the IEEE international conference on information and communication technology convergence (ICTC), pp 18–20
57. Ahmad I et al (2013) Uplink power control for interference mitigation based on users priority in two-tier femtocell network. In: Proceedings of the IEEE international conference on information and communication technology convergence (ICTC), Jeju, Korea, pp 474–475
58. Ahmad I, Chen W, Chang KH (2016) Co-channel interference analysis using cooperative communication schemes for the coexistence of PS-LTE and LTE-R networks. In Proceedings of the Vietnam, IEEE ICCE special session on LTE technologies and services, pp 181–182
59. Xiao J et al (2018) eMBMS and V2V communications for vehicle platooning in eV2X system. In: Proceedings of the conference of Korean institute of communications and information sciences (KICS)
60. Mughal UA et al (2019) Virtual cells operation for 5G V2X communications. In: Proceedings of the conference of Korean institute of communications and information sciences (KICS), Korea, pp 1–2
61. Mughal UA et al (2022) UAVs path planning by particle swarm optimization based on visual-SLAM algorithm. Intelligent unmanned air vehicles communications for public safety networks. Springer Nature Singapore, Singapore, pp 169–197
62. Chen W et al (2016) Analysis on the co-channel interference for the coexistence of PS-LTE and LTE-R networks. In: Proceedings of the conference of Korean institute of communication and information sciences (KICS), Jeju, Korea, pp 202–203
63. Ahmad I et al (2013) Block error rate and UE throughput performance evaluation using LLS and SLS in 3GPP LTE downlink. In: Proceedings of the Yongpyong resort, Korean institute of communications and information sciences (KICS), Daegwallyeong-myeon, South Korea, pp 512–516
64. Ahmad I, Kaleem Z, Chang KH (2014) QoS priority based femtocell user power control for interference mitigation in 3GPP LTE-A HetNet. J Korean Inst Commun Inf Sci 39B(2):61–74

# Subject Index




© The Editor(s) (if applicable) and The Author(s), under exclusive license to Springer Nature Singapore Pte Ltd. 2026



I. Ahmad et al. (eds.), *Reconfigurable Intelligent Surfaces for Unmanned Aerial Vehicle Communication*, Unmanned System Technologies, https://doi.org/10.1007/978-981-96-9662-8

**U**

# Author Index



© The Editor(s) (if applicable) and The Author(s), under exclusive license
to Springer Nature Singapore Pte Ltd. 2026
I. Ahmad et al. (eds.), *Reconfigurable Intelligent Surfaces for Unmanned Aerial Vehicle Communication*, Unmanned System Technologies,
https://doi.org/10.1007/978-981-96-9662-8

The manufacturer's authorised representative in the EU is Springer Nature Customer Service Centre GmbH, Europaplatz 3, 69115 Heidelberg, Germany. If you have any concerns regarding our products, please contact ProductSafety@springernature.com

Printed and bound by CPI Group (UK) Ltd, Croydon, CR0 4YY
29/09/2026
02224599-0004